# Tourism
in European Cities

# Tourism in European Cities

## The Visitor Experience of Architecture, Urban Spaces and City Attractions

John Ebejer

ROWMAN & LITTLEFIELD
*Lanham • Boulder • New York • London*

Published by Rowman & Littlefield
An imprint of The Rowman & Littlefield Publishing Group, Inc.
4501 Forbes Boulevard, Suite 200, Lanham, Maryland 20706
www.rowman.com

6 Tinworth Street, London, SE11 5AL, United Kingdom

Copyright © 2021 by John Ebejer

*All rights reserved.* No part of this book may be reproduced in any form or by any electronic or mechanical means, including information storage and retrieval systems, without written permission from the publisher, except by a reviewer who may quote passages in a review.

British Library Cataloguing in Publication Information Available

**Library of Congress Cataloging-in-Publication Data**

Names: Ebejer, John, author.
Title: Tourism in European cities : the visitor experience of architecture, urban spaces and city attractions / John Ebejer.
Description: Lanham, Maryland : Rowman & Littlefield Publishing Group, [2021] | Includes bibliographical references and index. | Summary: "This book explores urban tourism with a particular focus on European cities"—Provided by publisher.
Identifiers: LCCN 2021018212 (print) | LCCN 2021018213 (ebook) | ISBN 9781538160541 (cloth) | ISBN 9781538160770 (paperback) | ISBN 9781538160558 (epub)
Subjects: LCSH: Tourism—Europe—Planning. | Architecture—Europe. | Urban tourism—Europe. | Tourism—Economic aspects—Europe.
Classification: LCC G155.E8 E24 2021 (print) | LCC G155.E8 (ebook) | DDC 914.0456/2—dc23
LC record available at https://lccn.loc.gov/2021018212
LC ebook record available at https://lccn.loc.gov/2021018213

*This book is dedicated to the memory of*
*Edwin Ebejer (1954–2019)*
*Mario Mallia (1961–2020)*
*Victor Abdilla (1950–2020)*

# Contents

List of Case Studies     xi

List of Figures     xiii

Foreword     xv

Preface     xvii

**1 Introduction**     1
    1.1 Researching urban tourism     1
    1.2 Why do people visit cities?     8

**2 Changing Cities**     11
    2.1 The development of cities     11
    2.2 Competing cities and place marketing     13
    2.3 Place marketing and tourism product development     18
    2.4 The development process and urban planning     22
    2.5 Urban regeneration     26

**3 The Changing Dynamics of Tourism and Leisure**     37
    3.1 Changing tourism practices     37
    3.2 The changing role of culture in tourism     38
    3.3 Blurring of the boundaries between city residents and tourists     39
    3.4 Improved connectivity between cities     41

**4 City Resources for Tourism (1): Areas in the City**     47
    4.1 The tourism product     47
    4.2 Tourism districts and leisure settings     54

| | | |
|---|---|---|
| **5** | **City Resources for Tourism (2): Visitor Attractions** | **65** |
| | 5.1 Museums and sites of cultural significance | 68 |
| | 5.2 Cathedrals, churches and religious buildings | 79 |
| | 5.3 Nature-based attractions | 83 |
| | 5.4 Theme and amusement parks | 88 |
| **6** | **City Resources for Tourism (3): Accommodation and Other Facilities** | **91** |
| | 6.1 Tourism accommodation | 91 |
| | 6.2 Food and drink establishments | 99 |
| | 6.3 Facilities for shopping | 100 |
| | 6.4 Venues for conferences and exhibitions | 102 |
| **7** | **Walking and the Tourist Experience** | **107** |
| | 7.1 The nature of walking | 107 |
| | 7.2 Tourist motivations and walking | 111 |
| | 7.3 Walking the city | 113 |
| | 7.4 Pedestrian areas in city centres | 119 |
| | 7.5 Photography and the tourist experience | 127 |
| **8** | **Meaning of Place and the Tourist Experience** | **131** |
| | 8.1 The experience of place | 131 |
| | 8.2 How buildings and spaces acquire meaning | 135 |
| | 8.3 Tourist experience and the search for meaning | 137 |
| | 8.4 The role of narrative and the past in the tourist experience | 139 |
| | 8.5 Tourists as performers | 141 |
| **9** | **Urban Conservation and the Tourist-Historic City** | **145** |
| | 9.1 Urban heritage and tourism | 145 |
| | 9.2 The conservation of historic areas | 147 |
| | 9.3 Funding and the role of public authorities in urban conservation | 152 |
| | 9.4 World Heritage Sites and historic cities | 158 |
| **10** | **Architecture and Tourism** | **165** |
| | 10.1 The role of architecture in city tourism | 165 |
| | 10.2 Museum architecture | 168 |
| | 10.3 Iconic architecture and the city | 172 |
| **11** | **Cities and Events** | **181** |
| | 11.1 What are events? | 181 |
| | 11.2 Festivals and city spaces | 184 |
| | 11.3 Why cities choose to be eventful | 187 |
| | 11.4 European Capital of Culture | 193 |

| | | |
|---|---|---|
| **12** | Post-Pandemic Prospects: Overtourism or 'Undertourism'? | 199 |
| **13** | Conclusion | 207 |

| | |
|---|---|
| Appendices | 211 |
| **A**    World Heritage Site Criteria for Cultural Properties | 213 |
| **B**    Historic City Areas Inscribed as World Heritage Sites | 215 |
| Notes | 229 |
| Bibliography | 261 |
| Index | 277 |
| About the Author | 285 |

# List of Case Studies

| | | |
|---|---|---|
| CC1 | The regeneration of Lisbon's city centre | 31 |
| CC2 | Regenerating Sheffield | 33 |
| CD1 | The reconstruction of Rotterdam Centraal | 43 |
| CD2 | Billund Airport, Denmark, and the effects of low-cost airlines | 44 |
| AR1 | Three examples of the creative use of water in leisure areas | 59 |
| AR2 | Liverpool's waterfront development | 62 |
| VA1 | The Black Country Living Museum, near Birmingham | 75 |
| VA2 | The Guinness Storehouse, Dublin | 76 |
| VA3 | Churches in Salzburg Old Town | 82 |
| VA4 | Sheffield Winter Gardens | 85 |
| VA5 | Stockholm's National City Park | 85 |
| WK1 | Bankside and the Millennium Bridge, London | 116 |
| WK2 | Walkable urban spaces in Barcelona | 125 |
| UC1 | Restoration and adaptive reuse of fortifications in Malta | 156 |
| UC2 | Suomenlinna, Helsinki: Conservation and adaptive reuse | 162 |
| AT1 | The redesign of the Military History Museum, Dresden | 171 |
| AT2 | Centre Georges Pompidou, Paris | 176 |
| AT3 | Guggenheim Museum, Bilbao | 177 |
| AT4 | Norwegian National Opera and Ballet House, Oslo | 178 |
| AT5 | The Temppeliaukio Church, Helsinki | 179 |
| EV1 | Il Palio di Siena | 186 |
| EV2 | Galway Arts Festival | 192 |
| EV3 | Stockholm Culture Festival | 193 |
| EV4 | Glasgow 1990, European Capital of Culture | 196 |
| EV5 | Lille 2004, European Capital of Culture | 197 |
| EV6 | Valletta 2018, European Capital of Culture | 197 |

# List of Figures

| | | |
|---|---|---|
| Figure 1.1 | A framework for an urban tourism research agenda | 4 |
| Figure 3.1 | Rotterdam Centraal | 45 |
| Figure 4.1 | Functional segmentation of city facilities and resources | 53 |
| Figure 4.2 | Views of the historic area of Ostrów Tumski, Wroclaw, from across the river | 58 |
| Figure 5.1 | The most visited museums in Europe in 2019 | 71 |
| Figure 5.2 | The beautifully adorned loggia of the Castello del Buonconsiglio, Trento, Italy | 72 |
| Figure 5.3 | Sheffield Winter Gardens | 86 |
| Figure 6.1 | Tourist accommodation as a product | 92 |
| Figure 7.1 | A visitor to the Acropolis in Athens needs to do a lot of walking to fully appreciate the heritage and aesthetic values of this important site | 111 |
| Figure 7.2 | The Millennium Bridge is a popular pedestrian bridge across the River Thames, London | 117 |
| Figure 7.3 | A popular pedestrian urban space alongside a historic canal at Nyhavn, Copenhagen | 123 |
| Figure 7.4 | Rambla de Mar, a pedestrian walkway on timber decking at Barcelona | 126 |
| Figure 7.5 | Being photographed alongside the Little Mermaid in Copenhagen | 129 |
| Figure 9.1 | Fort St. Elmo, Valletta | 158 |
| Figure 10.1 | Military History Museum, Dresden | 172 |

# Foreword

As Europe emerges from one of the most difficult periods in recent history, a year that has rendered the indoors our immediate – and in some cases – only environment, and a year that has narrowed our experience of the world immeasurably, the time has come for us to look forward to expanding our horizons again and to explore all that our countries and our continent have to offer. Reflecting its millennia-long and chequered history, Europe has long offered tourists diverse and enriching experiences. These range from wondrous strolls through cities that resemble open air museums to the more humbling encounters with prehistoric structures. Both are, incidentally, available in the author's and mine native Malta and Gozo.

The European Union itself is a crystallisation of this history and while tourism is mainly a competence of its member states, it most certainly has important roles to play in this field. The European Capital of Culture is now in its thirty-sixth year. It is a frequent invitation to Europeans to visit previously unknown corners of our continent and enhance their sense of being European. The European Regional Development Fund has funded untold numbers of projects dedicated to the preservation and restoration of sites of historical importance across Europe, making them accessible to the public without risking any compromise to their aesthetic and structure. And responding to recent discoveries of the damage that climate change, acid rain and other environmental factors have wreaked on structures thousands of years old, the European Parliament has supported making future funding streams available to tourism-related investments with in-built plans for the reduction of carbon emissions across the board.

European legislation also amplifies tourism experiences indirectly through better regulation of a number of environmental criteria. This ranges from the improved quality of bathing water and the iconic blue flags flying on many European beaches to cleaner air, and greener cities with more efficient vehicles driving through them. Crucially, it also ensures that our cities' buildings, public spaces and transportation are accessible to all citizens including those with disabilities, the elderly and families with young children. And through its flagship **ERASMUS+** programme, the European Union provides for better education and training for future generations of architects, engineers, urban planners and designers who will ultimately be the ones developing and conserving our built environments for decades to come.

Over the next years in the immediate post-pandemic future, with the reopening of our societies and our economies, European residents will be invited to become tourists once again, starting in the own countries. Tourism sustains tens of thousands of livelihoods across our Union and there will be direct economic benefits to be gained from the re-opening of the sector. Through the Recovery and Resilience Facility, the European Union is encouraging member states to propose recovery strategies for their countries that are forward looking, digital and innovative, and it will be backing up those plans with direct funding and loans coordinated at the European level. I look forward to the revolution that this will bring, also in the tourism sector. Beyond the economic considerations, this re-opening will be a chance to rediscover the simple joy of open spaces anew, especially in the hearts of our metropolises and urban areas that have been locked down for so long. More than ever before, Europeans will be looking for a holistic tourism experience that does not just offer a window to our past but also spaces to spend their time with friends and family, to pause and enjoy the present in clean, accessible and beautiful spaces in Europe.

It is the availability and provision of moments such as these, which bring us politicians most satisfaction that our work is yielding tangible results.

Dr. Roberta Metsola<br>
First Vice President of the European Parliament

# Preface

The purpose of this book is to help readers develop an understanding of the relationship between tourism activities and the built environment. *Tourism in European Cities* is a first in that it considers urban tourism within the context of European cities. Apart from giving the theory and explaining the concepts, the book uses examples and case studies from European cities to illustrate them.

How did this book come about? I have been teaching at tertiary level for eighteen years. Much of what I taught, and still teach, is about tourism in cities. Teaching at tertiary level requires preparation and a lot of reading. A substantial part of the content of this book is based on that reading. It is augmented with the doctorate research which I completed in 2015 as well as various journal articles and book chapters I wrote since then. Three years ago, I wrote the initial tentative draft of one of the chapters. Since then, I read and researched extensively to ensure that the book presents the subject to readers in the clearest and most effective manner possible.

In the years prior to the COVID-19 pandemic, I had the good fortune to visit several cities across Europe and this provided me with further information and inspiration.

In writing this book, I strove to broaden my research to as many European countries as possible. I felt this was essential to give readers a more comprehensive view of urban tourism in Europe. Tourism graduates and practitioners are highly mobile. Many are likely to take up work and establish residence in a place other than their university or home town.

As evidenced by the reference lists and the bibliography, this academic work has been extensively researched. Tourism is a discipline that borrows from various other disciplines. The book reflects this and is inspired and informed by academic literature from the fields of tourism, urban geography, urban planning and architecture.

The book adopts a 'need-to-know' approach. In writing it, I constantly kept in mind the question: What information and knowledge do tourism and urban geography students need to be aware of? The same question is applicable to practitioners and professionals working in the tourism, cultural and related sectors. Before becoming a full-time academic, I spent many years working professionally in tourism primarily on product development. Because of it, I am well placed to distinguish between what is essential tourism knowledge and what is not. I was careful to go no further than what I consider to be necessary.

The book is essential reading for students of tourism and urban geography. It is also of interest to students of urban planning and architecture and to anyone keen to learn more about tourism and about European cities.

I would like to acknowledge two renowned academics with whom I have collaborated on various academic initiatives: Prof John Tunbridge and Dr Andrew Smith. Both have published widely and are leading experts in their respective fields. Over the many years that I have known them, they were a source of inspiration and guidance and for that I am very grateful.

A special word of thanks to my wife, Joanne, for her constant support in this and in my various other endeavours. Also thanks to my son, Jonathan, and daughter, Rachel, for their support and assistance.

John Ebejer<br>
March 2021

*Chapter 1*

# Introduction

## 1.1 RESEARCHING URBAN TOURISM

The study of tourism is the study of people who are away from their normal environment. It is also the study of the facilities and services that tourists make use of and the impacts that come about from the tourism activity. It involves the motivations and experiences of the tourists, the expectations and adjustments made by the residents of the destination and the roles of agencies that intercede between them.[1] Tourism is an industry that manages and markets a variety of products and experiences to people who have a wide range of motivations, preferences and cultural backgrounds and who are involved in a dialectic engagement with the host community. The outcome of this engagement is a set of consequences for the tourist, the host community and the industry.[2] Until relatively recently participation in tourism was restricted to those who had the resources and time to travel. Higher incomes and greatly enhanced mobility have enabled more people to engage with tourism. Today the vast majority of people in the developed world are tourists at some point in their lives. Tourism is no longer the prerogative of the few but is now a normalised and expected part of the lifestyle of a large and growing number of people.

Defining urban tourism is not as straightforward as one would expect. Although urban tourism can be defined as tourism in towns and cities, it is apparent that the field is more problematic as urban tourism is a complex phenomenon that encompasses a wide range of activities. Adding the adjective 'urban' to the noun 'tourism' establishes the spatial context but does not define or delimit that activity.[3]

Smith[4] draws a distinction between 'tourism in cities' and 'urban tourism' arguing that the former is tourism that incidentally occurs in urban environments, whereas the latter refers to tourists that specifically visit a city to experience it. He argues that tourism in London is dominated by urban tourism as most people who visit 'experience the chaos and diversity of a postmodern metropolis'. People who visit a city for business or some other non-leisure reason will spend at least some time engaging in leisure activities, enjoying the sights and the experiences that the city has to offer. Non-leisure visitors to a city are therefore an integral part of the urban tourism phenomenon.

The environment in town and cities is commonly referred to as being 'urban'. A simplified description of 'urban' would be any area that contains groupings of buildings, roads, paved areas and other forms of human intervention, to the extent that the area is perceived to be a built-up area. Edwards et al.[5] offer a more comprehensive definition. They define urban area as a functional and physical environment with a web of social, cultural, political and economic relationships and interactions between many individuals and groups. Such an environment is characterised by the following:

- A strong and broad economic base that is serviced from multiple cores for major business and professional activities
- A significant public transport network that acts as a gateway to other areas
- A significant population with a workforce that commutes to and from the multiple cores
- Long-term planned development

As Ashworth and Tunbridge[6] rightly note: 'Cities are important to tourism and tourism is important to cities'. Urban environments across Europe and globally have for many years been among the most significant of most tourism destinations.[7] Yet, until fairly recently, researchers have dedicated a disproportionately small amount of attention to it. Generally speaking, those studying tourism neglected cities, while those studying cities neglected tourism. There were few research works that focused on the linkages between tourism and the built environment. At the time, the most notable and outstanding piece of work about this area of study was Gregory Ashworth and John Tunbridge's 'Tourist-Historic Cities',[8] first published in 1990.

In deriving a definition, it is useful to consider certain features of urban tourism that distinguish it from other forms of tourism:[9]

- Tourists visit cities for many purposes. Significant numbers of tourists in urban areas are visiting for a primary purpose other than leisure including business, conferences, shopping and visiting friends and relatives.

- Larger cities easily absorb large numbers of tourists to the extent that, in most parts of the cities, the tourism activity becomes indistinguishable from the normal daily city life.
- Tourists make an intensive use of many urban facilities and services but little of the city has been created specifically for tourist use.
- Tourism in an urban context is just one of many economic activities and it must compete with a number of other industries for resources such as labour and land. Consequently, tourism planning and policy-making processes are made more complex by the necessary engagement between tourism and other policy areas.
- Within cities, there is a complex mix of constraints on development including constraints relating to cultural heritage and good neighbourliness for residential areas.

It is widely acknowledged that tourism is a force for change and that it creates a wide range of economic, social and environmental impacts – both positive and negative.[10] The difficulty, however, is to differentiate between changes that have been brought about by tourism and those that are attributable to other economic and social forces that are unrelated to tourism. For some tourist developments, the impacts are clear. Some obvious examples include a new hotel development changing the rural landscape or a city skyline; the development of a theme park taking up large areas of land; the generation of employment in tourism and leisure and so on. In many cases, however, it is exceedingly difficult to isolate the principal causes of change. It may be difficult to determine whether the changes are directly attributable to tourist development or whether tourism is one of a number of agents of change.[11] The argument is applicable to all forms of tourism but it is possibly more directly relevant to urban tourism. The complex interactions of tourism phenomena, coupled with the complexities of the urban environment, makes the measurement of urban tourism's impacts almost impossible.

Research on urban tourism is essential because of the complexity of the elements listed earlier in conjunction with the potential economic, social and environmental effects that may result from visitation to cities. In light of this, there is a need for more strategic and cohesive research and hence the need of a research agenda. In a study on urban tourism, Edwards et al.[12] present a considered and interrelated approach for the development of a research agenda. In their study, they adopted a range of methodologies to consult with interested stakeholders, including in particular persons who are directly involved in tourism policy and management.

Study by Edwards et al. identified a very diverse range of micro issues. These were grouped into a series of conceptual sets, the four more important ones being (1) tourist experience and behaviour, (2) impacts, (3) tourism

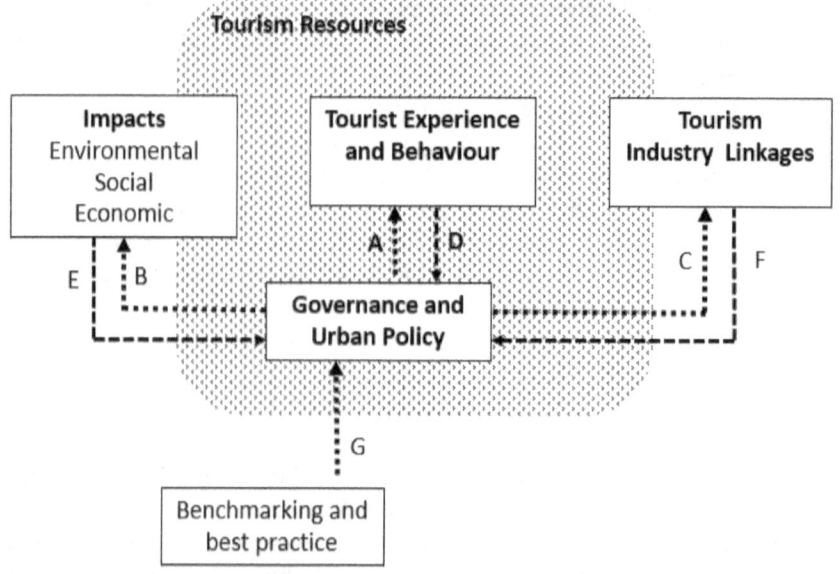

**Figure 1.1  A framework for an urban tourism research agenda.** *Source*: Adapted from Edwards et al., 2018.

industry linkages and (4) governance and urban policy. These are shown in figure 1.1. Tourism activity in urban environment is reliant on certain attributes, features and places within the city collectively referred to as tourism assets or resources. Also shown in figure 1.1 is a shaded rectangle representing resources, underlying the four conceptual sets listed earlier.

Developing a research agenda must recognise the manifold ways in which these resources are significant, namely the following:

- Experiences are structured around tourism resources.
- Excessive or inappropriate use can impact negatively on the resources themselves and on the environmental, social and economic context. Conversely resources can be used to generate positive outcomes for the community.
- The tourism industry relies upon resources as the basis for generating income.
- Good governance is needed to appropriately enhance, protect and manage the resources in the long-term public interest.

Tourism activity in cities is reliant on the infrastructure and the natural and built environments which are available for use by tourists and the industry within the destination. Essentially resources drive tourism within the city destination and represent the basis for generating income for tourism businesses.

Edwards et al. also consider transport access to and within the urban environment as one of the tourism resources. Different tourists will have their own perceptions of what the resources of an urban destination are. These may be either tangible or intangible. Competence is needed to recognise the resources and their respective values as well as what gives each of them that value. Such competence is required to appropriately develop and manage tourism in urban destinations.[13]

The research agenda being proposed seeks to provide appropriate guidance on the governance of city destinations and the specific resources within them. Governance incorporates urban policy, planning, design, management, place marketing and communication activities. Good governance for tourism necessitates a whole-of-government approach that aims to achieve positive outcomes with respect to:

(a) enhancing experiences for the tourists,
(b) reducing negative impacts and generate net benefits for the host community, and
(c) improving functioning of the interdependent businesses operating within the city.

The aforementioned relationships of 'governance and urban policy' are represented by arrows A, B and C, respectively, in figure 1.1. Improved governance of urban tourism destinations must be informed by a fundamental understanding of tourist experiences and behaviour, tourist impacts and also industry linkages – represented by arrows D, E and F, respectively, in figure 1.1. In essence, these various processes should revolve around the long-term maintenance and effective functioning of the resources upon which urban tourism is based. This requires an understanding of what those assets comprise, what gives them value, how that value can be maximised and what threatens to diminish their value.

Research focused on 'benchmarking and best practice' can further inform and guide urban planning and governance by pointing to practices that have produced successful outcomes in the past. This is represented by arrow G in figure 1.1.

Discussion on a research agenda for urban tourism is of interest not only to researchers and educators but also to tourism professional and practitioners. They are the people who derive policies and take decisions that determine the shape and form of urban tourism. Their decisions and actions greatly affect the tourist activity, the tourism industry and the context within which tourism operates. Such a discussion is also relevant to students as potential future operators in the tourism industry. Developing a research agenda provides appropriate guidance to the tourism industry on how to more effectively develop, manage and market urban tourism destinations. This is essential for the long-term sustainability of tourism in cities.

This book is intended for tourism professionals, practitioners and students. It is also intended for tourism operators with a business or geography background but who have never studied tourism. For them too, a basic understanding of the concepts discussed in this book is essential.

One of the purposes of this book is to contribute to the urban tourism research agenda, or at least to part of it. It is useful to consider where this book sits within the research framework described earlier. The earlier-mentioned framework talks about four conceptual sets (experience/behaviour, impacts, linkages and governance) with a fifth set, tourism resources, overlapping with all four. The focus of this book is primarily on tourism resources with chapters 4, 5 and 6 dedicated exclusively to them. Chapters 9, 10 and 11 deal with specific aspects of tourism resources namely urban heritage, contemporary architecture and events, respectively. An understanding of the tourist experience/behaviour is developed in chapters 7 and 8. Chapters 2 and 3 touch upon issues of governance and urban policy largely by debating cities as the context of urban tourism. Other than marginally in some chapters, the book does not debate tourism impacts largely because impacts would require several chapters to do it justice.

Most tourism takes place in cities. Of course, there are tourism and leisure activities that take place in non-urban areas, such as nature parks and skiing resorts, but even here tourists will make use of an urban area to travel to or to support their leisure engagements. This is being said without wanting to diminish the importance of tourism and recreation in natural areas. For many, the lure of the natural environment is essential for relaxation and enjoyment.

A common theme that runs through this book is the physicality of cities, that is, the architecture and the urban spaces. Tourism activities take place in buildings and in urban spaces within the city. The physicality of the city is expressed in its architecture. For many people, the word architecture brings to mind buildings that are outstanding because of their design, modernity or their historicity. A broader understanding of architecture is not limited to a few notable buildings but includes all buildings and also all urban spaces. In this book, architecture refers to the built environment of the city. This includes buildings, urban parks, roads, transport infrastructure and any other structures and spaces that are required for cities to function.

Urban design refers to the design and creation of spaces, normally in an urban context, which are amenable to people, including both residents and tourists. It is important to note that urban design is considered to be an integral part of architecture and hence its inclusion in any tertiary education programme for architects.

The following is an outline of the overall structure of the book with a brief description of each chapter.

- This chapter, as you have seen, sets the scene for the rest of the book. It describes the research agenda that is set out in tourism academic literature and notes where this book sits within that agenda. This chapter also discusses what motivates people to travel to cities and hence the reason that makes cities major centres of tourism activity.
- Cities are in constant change due to various social and economic forces. Cities seek to guide these changes in a manner that will enhance their competiveness in a global context of stiff competition between cities. Chapter 2 discusses these forces including the role place marketing and tourism have on how cities develop and evolve over time.
- Changes in cities are also brought about by tourism. Chapter 3 discusses the way the dynamics of the tourist activity have changed in the past three or four decades. Such a discussion is needed for a better understanding on how tourism changes cities.
- Chapters 4, 5 and 6 give an overview of the various resources that make up the tourism product. Chapter 4 explains tourism product and outlines how tourism district are an important part of that product. Chapters 5 and 6 consider the various facilities made use of by tourists in particular visitor attractions and tourism accommodation.
- Walking is an activity to which the tourist will dedicate significant time during a leisure visit to a city. For this reason, it merits a chapter in a book on urban tourism. Chapter 7 discusses the nature of walking and places the discussion within the context of the tourist experience. The chapter also discusses pedestrian areas in cities as these areas often become a focus of tourism activity.
- Chapter 8 develops an understanding of the tourist experience by investigating those elements that are most influential in shaping the experience at a city destination. The main focus is on the role of meaning and a sense of place in this process.
- In many cities, the historic core plays a crucial role in tourism. Apart from attractions located within it, the historic core is often an attraction in its own right with many tourists spending time walking and exploring the historic area. Chapter 9 takes a closer look at historic areas and considers the relationship between urban heritage and tourism. This chapter also looks at a number of cities that have been designated World Heritage Sites.
- Many cities provide a diversified architectural structure from different eras of their history including a broad range of contemporary architecture. Chapter 10 debates how new contemporary architecture developments, including iconic architecture, are used by cities to enhance their attractiveness in the

context of increased competition between cities. A section is also dedicated to museum architecture.
- Cities are increasingly resorting to events as a tool to deliver a wide range of outcomes and to reposition and differentiate the city from its competitors. Chapter 11 explains how events generate inward investment for the enhancement of public spaces and for infrastructure projects. The city often serves as a backdrop to the urban festival as activities and performances are staged in the city's urban spaces.
- Enter 2020 and with it the beginnings of the global COVID-19 pandemic. The pandemic has given rise to great uncertainty and brought with it difficult issues of unprecedented complexity. City destinations that previously suffered from overtourism are now faced with an opposite concern; after the effects of the pandemic have subsided there will be insufficient numbers of tourists to sustain their economies. Chapter 12 offers some reflections on the future prospects of tourism in 2022 and beyond.
- Chapter 13 offers some concluding remarks.

## 1.2 WHY DO PEOPLE VISIT CITIES?

Many city authorities that promote tourism are sometimes not fully aware why people visit. The links between the various motivations and the deeper reasons for visitation are not sufficiently appreciated.[14] The following gives an overview of the main reasons why visitors are attracted to cities. An awareness of tourists' motivations is a pre-requisite for a better understanding of urban tourism.

Discretionary free time is the time that is left over after taking into account work-related activities, daily chores and sleep. Many people dedicate significant amounts of their free time to leisure activities out of the home and this very often includes travel. Progressively over the past half-century, work practices have changed with a reduced working week and greater possibility to take early retirement. This has increased the amount of time available to people for travel thus facilitating the growth of tourism.[15]

There are many reasons why tourists are drawn to cities but one of the most important ones is its selection of visitor attractions. As illustrated in chapter 4, there is a great variety of attractions. Most leisure tourists do not visit a city for one attraction only but for a bundle or mix of attractions which meet their requirements and expectations.[16] In many medium-sized cities, the main attractions are clustered within one area of the city, normally the city centre, and are therefore within walking distance of each other. In larger cities, the selection of attractions stretches across different parts of the city. Prideaux[17] refers to cities as 'fun places'. They offer visitors the opportunity to participate in a multitude of activities and pleasures. Apart from visitor attractions, the activities and attractions offered by cities include dining, shopping, casinos, entertainment and nightlife districts and cultural districts.

The variety on offer has increased in recent decades as travel has become easier and cheaper and consequently more commonplace for a wider range of people. Overall, there has been a move towards shorter, more frequent and more tightly scheduled holidays spread throughout the year. The increase in short breaks is perhaps the factor which is most directly relevant to tourism in cities. Most visits to cities take the form of a short break of two or three nights rather than a long holiday.

There are several key attributes of cities that make them well suited to be tourism destinations.[18] Cities draw tourists to their attractions because these are often much better developed than in other types of destinations. They are easily accessible by means of rail or air travel. Most medium- and large-sized cities have a large stock of accommodation built to serve the business traveller. Cities offer all the essential services that tourists normally expect such as communications and transport infrastructure.

Prideuax[19] eloquently captures the essence of what makes cities important in the economic and cultural life of nations. The description given further is mostly relevant to capital cities and also to large and medium-sized cities. It is these same elements that make the city attractive to outsiders and hence are more reasons to visit:

> Cities are attractive spaces for visitors, the repository of commercial and national wealth and power, the creators and leaders of culture and fashion and the incubators of intellectual ferment. Cities also embody national spirit exhibited through the tapestry of public and private buildings that testify to the national origins and contemporary political and business power. They act as the custodians of national history and culture through institutions such as the church or other religious organisations, museums, art galleries and theatres. Popular culture is fermented in the creative precincts of city life and radiated out via the electronic media to the world.

Business travellers are an important component of travellers to cities. Depending on the city, the share of total visitors may be anything between 10 and 30 per cent. They make, however, a significant contribution to a city's income as they are high spenders.[20] Included in the business traveller category are delegates to conferences and exhibitions. The concentration of commercial, financial and industrial services in urban areas acts as a focus for different people to visit cities for employment-related purposes such as conferences, exhibitions and business meetings. Business people travel to meet with other business people for a variety of reasons including to promote goods and services, to discuss sales deals, to attend company management meetings, to provide consultancy services, to meet with professional consultants, to deliver short training courses and so on. Despite improvements in online communications, there still is the need for business people to travel and meet in person, although the 2020 pandemic might lessen that.

Another reason for people to travel to a city is to visit friends and relatives (VFR visitors). Cities are places with large populations and hence there is a high propensity for travel motivated by visits to friends and relatives. Many of the visits have a secondary motivation of enjoying a holiday and spending some leisure time away from home. The number of VFR visitors to a city is also influenced by the attractiveness of the place as some visitors will combine visiting friends with sightseeing.[21] Most VFR visitors do not involve the use of serviced accommodation and hence the overall expenditure may be less than other forms of tourism. On the other hand, the friend's familiarity with the city enables that VFR tourist to engage in activities that other tourists might not normally engage in such as going to the theatre or having a drink at the local pub. In the past two or three decades, there have been significant increases in VFR travel largely because of the increased number of people across Europe who relocate to other cities for reasons of work or study. Civic pride of city residents is a means for promoting VFR tourism as they invite friends and family to visit and are proud to show them around their city.

Tourism in cities is characterised by the variety of activities, experiences and facilities that is offered to visitors. This sets it apart from other forms of tourism such as resort-based travel in coastal or alpine contexts.[22] Kolb[23] argues that increasingly tourists wish to visit sites that reflect the daily cultural life of the local community. Even if tourists visit the well-known landmarks, more and more visitors are eager to experience and explore the daily life. Tourist looking for authentic experiences would wish to see, for example, local community members engaging in arts and crafts tied to a different time or culture. Examples include lace making and weaving.

For contemporary urban tourism, researchers have abandoned the idea of a unitary role type of the tourist. Current tourism discourse reflects a greater sensitivity towards traveller's differences in terms of demographic and lifestyle characteristics as well as interests, preferences and behaviour.[24]

Some travellers are familiar with a city's conventional attractions and so they deliberately seek places, amenities and experiences in other areas of the city. As a result, tourists have become a common presence away from the more popular tourism areas, prompting new opportunities and challenges for the city and its residents.[25] This also brings about significant changes in the leisure and tourism landscape of the city.

*Chapter 2*

# Changing Cities

## 2.1 THE DEVELOPMENT OF CITIES

Cities are as old as civilisation. Most major cities of the modern world have long histories as centres of government, religion and trade. The first urbanised settlements probably emerged in Neolithic times in the Middle East on the fertile river plains of Tigris and Euphrates Rivers. Small settlements grew into cities the earliest of which date back to 9,000 B.C. Trade and the need for defence were the circumstances that made the establishment of cities possible. Farmers living in fertile river plains were able to produce surplus agricultural produce. This underpinned the rise of an urban population who traded their output of manufactured goods, such as cloth, pottery and other household goods for the farmer's grains, animal products and other foodstuffs. Early cities usually emerged where the surrounding agricultural land could produce sufficient surplus to maintain non-food producing urban populations. These cities grew when they engaged in trade; first within the city itself and then between cities. Although the first cities came about in the Middle East, they soon emerged in other parts of the world including India and China and later in Africa and Europe. Some of these cities became fairly large. For example, Rome was estimated to have a population of 1 million towards the end of the first century B.C. In all cases, the key to sustaining urban life was the trade between the city and surrounding rural areas and nearby cities.[1] From the second century A.D. onwards, cities continued to evolve and develop as trade energised economic activity and cities assumed new responsibilities as centres of culture, scientific discovery, commerce and learning.

In history, changes to cities often came about as a result of economic changes and technological developments. In the middle of the nineteenth century, steam-driven factories and city-centre docks encouraged the establishment of densely concentrated settlements.[2] In many countries across Europe, particularly Britain, urbanisation was a product of industrialisation with vast numbers of people moving from the countryside into cities to provide labour in factories. The industrial revolution created and sustained urbanisation that in turn funded the growing wealth of main European cities. Wealthy rulers across Europe spearheaded ambitious projects to modernise European capitals in the second half of the nineteenth century.[3] Cities and their respective rulers wanted to demonstrate their greatness and progressiveness through the construction of magnificent buildings, monuments and parks. Large-scale urban remodelling schemes took place in Paris, Vienna, Barcelona and other cities. Cities also wanted to show themselves as leading players in industrial activity and production. For example, London's Crystal Palace was built as the centrepiece of the Great Exhibition of 1851. It was the cutting edge of architecture and building construction of the time though sadly this fine example of nineteenth-century architecture was destroyed by fire. In Paris, the 1889 Exposition Universelle was held to celebrate the hundredth anniversary of the storming of Bastille in the French Revolution. It included the engineering marvel of the Eiffel Tower, a structure that continues to enchant visitors to this day.[4]

Technological development, industrialisation and modernisation brought about the need for urban planning and urban expansion of European cities. These economic and social changes created waves of immigration from the rural areas into cities.[5,6] No urban area is static – every urban area is subject to change. Towns and cities across Europe underwent dramatic transformations since the mid-twentieth century. After the post-war boom, the mid-1970s was a time of recession. Many cities across Europe were faced with a recession for the first time in a generation. The recession and increased competition accelerated processes of industrial restructuring in many cities. Hardest hit were cities whose economies were dominated by older industrial sectors such as mining, steel production, heavy engineering and textiles.[7] Several processes were at work driving and shaping changes in cities.[8] Post-industrialisation saw a decline in production and in manufacturing jobs. Concurrently there was an increase in consumption and in knowledge services such as finance, insurance, retailing, information and creative industries.

Cities had always been focused on the city centre and the coming of the railways reinforced this. The railway station was a node of activity for both people and goods and as a consequence the various uses within the city were structured around it with shops and offices huddled close to the train station to service people from a wider area. The warehousing and manufacturing industries were also located within easy reach of the train station. Further out were the residential areas. As various uses developed ad hoc around the train station and beyond, the tendency was towards segregation of uses (e.g. keeping warehousing and industry separate from residential) but this did not always happen that way. The geomorphology of the land often determined how a city evolved with development being pulled towards particular areas such as river valleys. In the past sixty years, cities have witnessed significant changes in their internal geography. From being relatively compact and monocentric, cities have decentralised and become polycentric with consequences for core areas.[9,10]

In the 1970s and 1980s, many cities across Europe experienced a downturn in their manufacturing sectors. Industrial zones in many cities became derelict and unemployment rates were high. Cities were facing a complex set of challenges over the past decades and this has stimulated the development of urban policies oriented towards the strengthening of the competitiveness of cities.[11] The structural economic changes that affected European economies since 1970s have changed the nature and intensity of how cities approach urban development.

Concurrent with the downturn in manufacturing, debates on the revival of city-centre economies revolved around retailing and offices. It was expected that these activities would expand both economically and geographically and convert adjoining residential areas into commercial developments. These activities remained buoyant but did not expand in a way that was generally expected. Instead, by the late 1990s, the expanding city-centre functions were leisure and residential. These two are not unconnected as many of the new residents are moving back to the city centre because of the leisure amenities available, thus stimulating demand which results in the further development of facilities. In the space of twenty to thirty years, the character of the city centre has significantly changed with leisure becoming the main theme. Other than shopping for essential food and goods, retailing is perceived to be a leisure activity. The city centre caters for this type of shopping. Elsewhere there has been an expansion of visitor attractions, bars, restaurants, theatres, cinemas and concert halls, as well as hotels.[12]

## 2.2 COMPETING CITIES AND PLACE MARKETING

The shape and form of any city is the outcome of decades, even centuries, of economic, social, political and cultural forces that influence decisions that eventually shape the city. Cities of today face two choices. Either they develop to meet the challenges created by the pace of global change or they resist change and stagnate. Cities compete in a global economy. To remain competitive, cities adopt strategies that make use of their own innate resources – their histories, spaces, creative energy and talents. They constantly seek to redefine their economic roles as old functions are lost and new functions are sought to take their place. The forces driving economic activity in cities have changed over time; first from extractive industries to manufacturing, then on to services and now to the production of experiences. The functionalities of cities saw a similar progression from centres of production to service centres to backdrops for experiences.[13] While the lost functions are in the goods-manufacturing and goods-handling sectors, the new functions involve the creation, exchange and use of information. Many towns and cities have repositioned their economic and employment structure to develop the advanced service sectors activities such as banking, finance, business services, corporate and public sector employment as well as the expansion and development of tourism and hospitality sectors.

Urbanisation was a major force that contributed to the development of European cities, from the nineteenth century right up to the latter part of the twentieth century. Primarily for work-related reasons, people moved away from rural areas into cities fuelling their growth. Populations engage in a diverse range of social, cultural and economic activities in the context of towns and cities. Such activities include tourism, leisure and entertainment thus making cities centres of service activities.[14]

Globalisation is a potent process of change. It intensifies worldwide social and economic relationships that link distant localities in such a way that local happenings are shaped by decisions and events that occur many miles away. Protection from competition has been eroded due to technological changes and communication improvements coupled with new liberal enthusiasm for deregulation and economic integration.[15] Cultural globalisation saw symbols and values spread across the world – leading to concerns about the erosion of local and regional identities and the homogenisation of the urban landscape.

Another process of change was new forms of urban governance. The way cities are governed has undergone significant changes since the 1980s. Traditionally, city authorities perceived their role to be 'managerial' and dealing solely with the delivery of services to city residents. This shifted to an 'entrepreneurial' role of city authorities, taking action to improve the city's competitiveness.[16] Authorities have shifted their core focus from the betterment of local conditions to outward-oriented policies designed to attract investment, tourists or new residents.[17] Today cities compete internationally to attract investors, corporations and talented people or to organise sport and mega-events. The entrepreneurial spirit of cities often involves the transformation of the urban environment and its spectacularisation.[18] The need for economic restructuring has made cities use cultural resources to create economic, social and cultural benefits. The creation and promotion of events such as festivals, exhibitions and fairs have become a crucial component of urban development strategy in many cities. These changes happened and continue to happen in a context of fiscal problems, economic restructuring and policy decisions at national and EU levels.

Cities have a wide range of functions and serve a broad range of aspirations. They are typically seen as engines of modern economic life. At the start of this century, there was a renewed optimism and a shift in perception. Once seen as problematic residues of past industrial economies, cities were being considered as exciting and creative places to live and work.[19]

The expansion of tourism-related activities has had an effect on the physical environment. From waterfronts to decaying industrial areas close to city centres, tourism and leisure projects have been used to anchor regeneration projects. New buildings have been constructed and old buildings brought back into use and the physical environment surrounding them enhanced. For an iconic development or for a major regeneration project of an area, there tends to be spillover effects on the surrounding areas as other activities are attracted in the form of hotels, restaurants and shops.[20]

Every city is unique. Urban areas are different when considered in terms of size, location, function, appearance and heritage. No two city destinations are identical and yet they have a common denominator – tourism. Many of the qualities that attract tourists are frequently similar among competing cities.[21] All cities are multifunctional – some more than others. Cities offer a range of facilities and attractions including historic urban spaces, monumental architecture, museums, religious buildings, visitor attractions, pedestrian shopping streets, tourist accommodation, culinary experiences, evening entertainment and so on.

Though different, there are certain qualities and features of city destinations that are similar. The buzz and vitality of a main shopping street in Dublin, for example, is not very different from that of a street in Vienna or Munich. The overall experience of a modern art museum in Amsterdam is not very different from that of a similar museum in Barcelona. A cathedral in Kraków has its own interesting features and narratives as does the cathedral in York or in Strasbourg, but in terms of the enjoyment of heritage, there are many similarities (an engaging narrative, visually interesting features, past building construction techniques, etc.). Most cities offer a good range of restaurants and hotels. Other than outstanding attractions, there has been a growing trend for cities to be perceived as similar by potential tourists. For those who have never visited, it will be very difficult to distinguish between two cities that have similar offerings and decide which of the two will offer the more enjoyable and engaging tourist experience.

Changes that happened in the past thirty years have increased the number of cities in Europe which have more or less similar tourism characteristics. The main factors that brought about this increase were the fall of the Berlin wall in 1989, the continued decline of major industrial production companies, the increased competition between cities and the increased integration of economies across the EU. It is very difficult, therefore, for a generic city with a range of suitable, but not outstanding, attractions to compete in an increasingly crowded market place. Place marketing of a city destination should avoid giving the impression that visiting will be 'more of the same'. It needs to offer something that distinguishes the city from its competitors.

The 1990s was a time when markets became more globalised. This was when the strategic use of place marketing techniques became more widespread.[22] It was also a time when cities increasingly became major economic entities on an international stage as national governments were less willing to finance their economic development. As deindustrialisation intensified, cities turned to campaigns of capital-intensive place marketing to transform and sell their cities as post-industrial centres for services, leisure and consumption. The shift was facilitated by the emergence of new media technologies and the increased professionalism of the marketing industry.[23]

Cities compete with one another regionally, nationally and internationally. They do so to attract tourists to fill their hotels, conference centres, stadia and museums. Their competitive advantage lies in their perceived appeal as places for play.[24] The efforts to promote and reposition the image of cities have translated into place marketing practices for the purpose of placing them on the European and global market. Place marketing concerns the way the city competes with other localities to attract consumers by creating and communicating narratives about its attractions.[25] A major aim is to attract spatially mobile capital by differentiating a particular city from its competitors. It has been increasingly recognised that an attractive place to visit tends to be an attractive place to live and work. Tourism therefore has become an important component of economic development.[26] Place marketing is not an entirely recent phenomenon and originated in industrial times when seaside resorts in Britain and across Europe sought to attract the working classes to visit by means of the railways.[27] The notion of place marketing is based on the assumption that the 'city' can be compared to an organisation in charge of satisfying the needs of particular customers and target groups. It also assumes that the 'city' is a commodity that can be packaged, marketed and sold. A new breed of 'destination branding managers' and 'place marketers' began to advise urban leaders about appropriate actions and strategies to increase a city's attractiveness.[28]

In professional and academic literature, the two terms 'place marketing' and 'place branding' are often used interchangeably.[29] In a recent study[30] on place marketing practices in Dutch municipalities, it was found that practitioners used the term 'place marketing' synonymously with place branding and place promotion when discussing tasks and responsible organisations. Boisen et al.[31] argue that there is a conceptual confusion regarding the use of the terms 'place marketing', 'place branding' and 'place promotion' and that this is rooted in the complexities of the diverse and constantly changing practices of cities to improve their competitiveness. They note that place promotion is limited to increasing the attention for what the places have to offer. In contrast, place marketing involves a more integrated approach and is predominantly concerned with fine-tuning the place to manage supply and demand, both through promotional measures and other measures aimed at improving the product-market combinations.

Another distinction that needs to be made is that between place marketing and mere place selling. Place selling is the communication of particular characteristics of a place with the help of logos, slogans, advertising campaigns and public relations exercises.[32] The role of the marketer is to identify, anticipate and satisfy the consumer's requirements and to do so in a profitable manner.[33] Even if they claim to be promoting and marketing, many destinations are merely in the process of place selling. Some cities go beyond selling, and adopt a place marketing approach that also addresses the product. For place marketing, efforts are made to enhance the products for tourist consumption, even if the marketeer has limited control over the actions and investments of the various departments of city authorities.

The marketing of a city destination is very different from the marketing of a normal product. Any other product can be seen, tested and compared with other similar products. For cities, it is the consumer, and not the product, that moves with the product being sold before the consumer actually sees it. This makes place marketing of crucial importance as one can have a great city but unless that is recognised by potential buyers, the number of visitors will be limited. Just as the reputation of commercial companies may rise and fall with the quality of their products and effectiveness of their marketing, so too may a city's reputation fluctuate according to local conditions (e.g. infrastructure, services and safety) and the effectiveness of the city's place marketing efforts.[34]

In marketing, the product and its benefits are matched with the needs and desires of the consumer. This matching is easier for a business as it can alter its product or even create a new product to meet consumer needs. Of course, the process of developing a city as a tourist destination follows a different sequence as the product already exists. The architecture, urban spaces, parks, scenery, history and ethnic heritage of the residents are already in place. These features can be enhanced but they cannot be fundamentally changed. Businesses are able to research the consumer before developing a product. In contrast, tourism marketeers first analyse the city's existing features and the benefits and then identify potential tourists who will be interested in what the city has to offer.[35]

A further complication in the marketing of cities is that different segments of potential tourists will want different features and benefits from the tourist experience. For instance, for travellers on a budget, reasonable prices in the city and the availability of low-cost accommodation are important considerations when deciding upon a travel destination. A family with children will seek out cities that have attractions that are enjoyable to both children and adults. There are many people who are keen on walking and hiking. For this group of people, a town surrounded by extensive countryside and having a network of nature trails would be very attractive. Because potential tourist segments will need and desire different benefits from visiting a city, marketeers must first carry out a product analysis of what the city has to offer. Only then will the tourism marketeer have the information needed to decide which potential tourist segment to target.[36]

## 2.3 PLACE MARKETING AND TOURISM PRODUCT DEVELOPMENT

According to Kolb,[37] marketing theory defines a product as consisting of one of three elements namely a tangible physical good, an intangible service or an experience. She argues that, in terms of being a product, a city is unique because it is composed of all three. The architecture, the parks, the streets and the monuments are the physical element, whereas services are provided by cultural facilities, visitor attractions, accommodation, catering and other service providers. The physical element and the service provide for the overall experience of the tourist at the city destination. Whatever the type of city destination, these three elements – physical, service and experience – provide the basis upon which the city's image and the marketing approach are developed.

To be effective, a place marketing plan has to be based on a proper understanding of the product. Marketing and product should not be divorced as no marketeer can draw up a proper marketing plan of a product about which they know little or nothing. Marketing is the process of matching the features and benefits of the product with the needs, desires and expectations of the consumer.[38]

Place marketing is a process by which place is associated with some desirable qualities that resonate with particular audiences. This can be achieved with targeted enhancements in the city's built fabric and urban spaces coupled with communication strategies that focus on select aspects of the city's local identity, history and culture.[39] Place marketing should not be detached from the actual development of the place.

To develop its tourism product, a city destination needs to do market research. On the demand side, it needs to understand what kind of experience the visitor would wish for. On the supply side, it needs to identify the tourism product opportunities and investigate their potential. It is then through a targeted market campaign that the tourism product (the supply) will be made known and promoted with potential visitors (the demand).

Integrating place marketing with the regeneration of the urban fabric provides for a more integrated approach to city development. The enhancement of public spaces and the urban fabric is an integral part of tourism product development which has the objective to upgrade the city environment and make it a pleasant place to live and visit. Place marketing normally falls within the responsibility of the city authorities as do city planning and management. There is an increasing understanding that this facilitates coordination and the development of an effective symbiotic relationship between these two activities.[40]

In developing a place marketing strategy, it is essential that the city is understood as a whole, and not merely as a piecemeal list of attractions.[41] In Germany, during the late 1980s, the notion of 'stadtmarketing' (city marketing) started to enjoy wide popularity among city authorities as they were facing increased economic and fiscal stress due to low growth and rising unemployment. By the mid-1990s, 60 per cent of German cities had explicit strategies of city marketing that were worked upon by specialised consultants or dedicated city marketing organisations.[42] In the Netherlands, in 2016, 80 per cent of municipalities had policies related to place marketing. About 32 per cent of municipalities had established organisational entities tasked with place marketing, branding or place promotion.[43]

When business people decide in which city they will invest, they inevitably consider economic factors such as labour supply, labour costs and property rent values. There are however also non-economic factors that they consider such as the lifestyle that the city offers to its personnel. The influence of lifestyle opportunities is likely to be greater if the firm has a high proportion of professional and managerial staff.[44] How the city is perceived is important in economic decision-making and hence the enhancement of a city's image is relevant not only for tourism but also to attract inward investment.

Apart from reframing the image of the city, place marketing boosts a spirit of renewal and reinvigorates the sense of pride and identity among city residents. A spirit of renewal also entails embellishment, upgrading and sometime rebuilding of the urban environment. Such efforts are designed to present cities as places to live in, invest in and visit.[45]

Place marketing was often used by cities as a managerial instrument to renovate the urban environment. This was achieved in one of two ways. Post-industrial cities facing decay sought to rebuild and reanimate cities with significant investments in the enhancement of urban spaces, provision of facilities and upgrading of the infrastructure. Bradford in the United Kingdom is one example of this approach.[46] Alternatively, cities re-planned their economies to become more service-oriented with a focus on cultural production as an urban survival strategy. Mega-events of sports and entertainment became powerful drivers for revenues. A good example is Barcelona in Spain which not only renovated its urban fabric for the Olympic Games in 1992 but also took that opportunity to enhance and promulgate the Catalan culture; promoting a vibrant image of the city and fostering the sense of pride among the residents.[47]

Cities are reorganising themselves by utilising cultural strategies to craft new identities, employing numerous initiatives aimed at growing their local economies. In response to fiscal pressures and other larger sociocultural changes, cities have aggressively embarked on plans to reinterpret and romanticise their past. They are also trying to replace former identities by constructing and promoting new culturally based images that rely on entertainment, leisure activities, urban tourism and conference tourism. The tools used to advance this new direction are various and include museums, festivals, revamped public spaces, ethnic precincts, conference centres and urban embellishment initiatives.[48]

The development of a viable tourism policy and a place marketing strategy necessitates an audit of the destination and its attributes; its strengths, weaknesses, problems, challenges and both past and current strategies.[49] An audit of the destination helps communicate information and issues to all parties engaged in policy formulation and is a key input to any effort to create and maintain a competitive destination. Effective use of the tourism product is essential to enhance a city's competitivity.[50] The competitive advantage of a city relates to its ability to use resources effectively over the long term. A city may have significant resources but is unable to use them to enhance its competitivity. On the other hand, a city might have limited resources but is able to maximise their tourism potential. The latter is more competitive and more likely to achieve success in tourism in the long run. Ritchie and Crouch[51] advocate five actions for city destinations:

(i) To have a tourism vision
(ii) To share the vision among all stakeholders
(iii) To understand its strengths and its weaknesses
(iv) To develop an appropriate marketing strategy
(v) To implement it successfully

They argue that a city that acts and operates in this manner will be significantly more competitive than one which has never asked what role tourism is to play in its economic and social development.

Several place marketing strategies have been used by cities with varying success. One possible strategy is to invest in contemporary architecture that will hopefully acquire an iconic status over time. Having it designed by an internationally renowned architect will make it more likely for the iconic building to become memorable to the potential visitor. Some iconic buildings are symbolic in that the main intention is for them to, hopefully, enhance the city's image to the outside world. Some others have a tourist related function, most commonly as a museum, over and above the symbolic value.[52]

Another approach for putting a city on the map is the staging of a major event. Organising a major international event, such as the Olympics, involves a significant expenditure for a city and indeed for the country. Some would argue, however, that even if the event is not financially successful in the short term the added value of a greatly enhanced international image will make the expense worthwhile.[53]

A further approach to place marketing is the enhancement of the public spaces especially those that are more relevant to tourism. These would normally be urban spaces within historic areas of the city but may also include spaces in the city centre or close to major visitor attractions. Public space enhancement would often involve pedestrianisation of some streets and hence the enlargement of the area within the city that is dedicated to pedestrians only.[54]

The identity of any nation is dynamic and unique, created by diverse features such as culture, history and location. According to Campelo,[55] 'National identity is a mosaic of many singular identities manifested by the different cultures of each part of the nation. As a dynamic entity, the nation provides a terrain where individual and collective selves interact and negotiate a supra identity where, potentially, everyone is connected'. Different cultures and ethnicities may result in internal tensions and create challenges in determining a national identity with which all groups can identify with. The same can be said for identities related to cities.

A city destination differs significantly from many of the goods and services for which modern-day marketing has been developed. Product marketing is – for the most part – the task of selling a product that is essentially indistinguishable from its competitors. Detergents, yogurts and life insurance schemes differ little from one another except in the packaging design but the particular brand one buys must be perceived to wash brighter, be tastier or provide greater future stability from its competitors. The urban tourism resource has more complex attributes when subjected to the known concepts of marketing. Unlike most products, the purchaser (the visitor) does not see or experience the product beforehand and is simply buying a desired satisfaction at an acceptable destination.

The concept of product development is difficult to apply to a complex urban environment. The urban cultural heritage encompasses not only the built heritage but also human activity. Both the cultural asset and the inherent depth of historic towns are being overlooked when, as a product, they are simplified to become historic fabric, street patterns and selected history. The urban totality cannot be portrayed to the visitor and expectations have to rely on the images and the words chosen to describe a place.

Many cities across Europe have developed positive images largely following many years, and even decades, of place marketing, combined with product development. Other cities may have the potential to develop a positive image in a manner similar to the more well-known tourist cities. They might not have done so due to limited resources or a political context that does not facilitate a proactive place marketing approach. There is a set of positive ideas that are associated with various cities including liveliness, excitement, richness of culture, cosmopolitanism and being full of opportunities. It is in cities that culture is created and displayed or performed at its highest levels in theatres, concert halls, art galleries, museums, covered arenas and stadia. There are many opportunities to meet other people and to socialise. The city offers many types of entertainment and spectacle, such as sports, sufficient to meet the needs of all tastes. In the city, there are opportunities to shop in diverse retail establishments including markets, small shops in historic areas and department stores. The city can also be a place of pleasant environments and townscapes and also of monumental buildings and contemporary architecture.[56] In short, the city is rich and diverse with many attractions for the tourist.

## 2.4 THE DEVELOPMENT PROCESS AND URBAN PLANNING

The quality of the urban environment affects the tourism experience. Even if the visitor attractions or historic areas are what motivates most tourists to visit, the aesthetics of the urban areas matter as they influence the overall perception of the city destination. One of the forces that shape the urban environment is the development process.

Property development is engaged in the production of commodities in the shape of various types of buildings developed to meet a demand from users. The buildings can be for various economic and social uses, including as residences, offices, retail outlets, tourism accommodation, industrial areas and transport infrastructure, among others.

Each new build involves a considerable investment of finance and effort. As new types of city functions appear and others die out, or as functions alter their locational or building requirements, these different changes become reflected in the built environment. The city expands laterally and adapts internally to the changing requirements of those requiring built space. Thus, urban change is driven by the dynamics of businesses and households, the outcomes being mediated through land and property markets.[57]

At the periphery of cities a property development typically occurs on a large scale occupying a sizeable area of previously uncommitted land. Peripheral developments are usually residential or industrial but there are also developments of shopping malls and office parks. Within the city, the scale of property developments is generally less extensive. Apart from new buildings and redevelopments, there are typically many refurbishments of existing buildings or their conversion to new uses. Although each site may be relatively small, the cumulative effect of numerous small-scale developments can be visually significant. In historic areas, landownership and developments were traditionally based on small individual plots. The bulk of interventions on the urban fabric are normally rehabilitation and extensions of existing buildings, although, in some cities, new buildings are not uncommon. For new developments in historic areas, the sheer scale may be incompatible with the small-scaled features of the historic built fabric and may thus be perceived to be intrusive. In whichever part of the city, property developments are regulated by a planning system the purpose of which is to guide change within the overall economic, social and environmental objectives of the city.

The fragmented nature of landownership in city centres and in other parts of the city means that large-scale redevelopment may require the assembly of a number of separate plots. During site assembly, developers will frequently operate surreptitiously, purchasing properties at arm's length perhaps using intermediaries. This is done to avoid having landowners inflating prices. A consequence of the length of time required for site assembly is the tendency to suspend building maintenance of those properties destined for demolition. Other delays may result from the planning process. The resultant dereliction can therefore be long term.

In many cities, redevelopment projects are carried out in areas of former industrial sites and disused docklands. The nature and extent of some of these projects are such that they fit into the city's overall regeneration strategy, a topic that is discussed further on in this chapter. Urban regeneration projects are often promoted and financially supported by the public sector through financial grants or tax inducements. If the land is publicly owned, the city authorities may incentivise development by transferring the land to the private sector at advantageous rates, provided that the development is carried out in line with the economic and social objectives of the city authorities for the area. Collaboration in this manner between public authorities and the private sector is referred to as public private partnership (PPP).

In the development process, there is a range of actors all acting within the institutional context and legal framework of the country. Private sector actors are driven by the profit motive. The process is initiated by the developer who perceives an opportunity for profit and takes action to put the development scheme together. They negotiate with the landowners to acquire the development rights for the site and arrange for short-term financing for site acquisition and to fund construction. Numerous sources of funds exist but the most common are short-term loans from the banks. Developers engage architects, engineers and planning consultants to design and devise a development scheme within specific cost constraints and that will be acceptable to the planning authorities, while still providing them with a profit. Developers also engage the builders and other contractors to carry out the construction up to completion. They then appoint estate agents to seek suitable tenants or purchasers for the completed building. In many instances, developers use the income from the sale of the completed building to pay off the loans and interests, leaving them with a profit in return for their efforts. It is also the developer who carries the risk of making a loss, as things can go wrong. For example, there could be a sudden downturn in the market resulting in reduced income from the sale of the property. The shape and form of the built environment in a city is the cumulative total of the actions and decisions of a large number of developers, taken over many years within the planning and legal framework.

In the closing decades of the twentieth century, the international trend towards financial liberalisation and the increasingly global operations of finance capital provided developers with a far wider range of credit sources than previously. Together with a favourable macroeconomic climate, this created more competitive borrowing terms for developers, thus facilitating a global property boom.[58]

Maclaran[59] defines urban planning as the strategic regulation of physical development as a means of controlling the production and use of urban space. The development processes, briefly described earlier, are influenced by urban planning processes, as are land and property rights and property markets. Urban planning is a form of collective action undertaken by the authorities in an attempt to influence, shape and control the outcome of property development initiatives. Planning normally intervenes on the development process by means of passive controls. There are instances, however, where authorities use the planning process to intervene proactively by means of financial or other incentives to encourage development outcomes that authorities deem desirable.

Land is a resource that is used for a wide variety of human activities such as residential, leisure, commercial and other uses. Tourism accommodation, visitor attractions, catering establishments and theme parks are some of many tourism-related facilities that make use of land. Different land uses have different requirements and impacts. An urban planning system is needed to ensure that the requirements of different uses are met while at the same time ensuring that the impact of a development on nearby properties and users are within acceptable limits. Urban planning aims at the efficient use of resources and that includes the efficient use of space. Resources are limited and therefore the process of planning aims to derive the maximum benefits for the community from available resources. Planning is also a means for identifying development opportunities and taking action so that such opportunities will come to fruition.

Planning is the organisation of future action to achieve defined objectives. Planning should be viewed in the context of anticipating change and be concerned with the future implications of current decisions. To plan well, you need to have some idea about future situations and issues.

Up to a few decades ago most planning systems in Europe were based on the preparation of a master-plan which was assumed to be sufficient for guiding and controlling future development patterns. It was found, however, that such plans were too rigid and do not take into account changing lifestyles, technology and other circumstances. The predominant approach nowadays is to have written policy documents that are supported with maps. Planning policies give a clear indication as to how development is to take place but at the same time allows some level of flexibility and interpretation to cater for the specific circumstances of each site. This planning approach also provides sufficient flexibility for amendments and improvements to be made to policies in line with changing circumstances or new information.

Most European planning systems give significant importance to the involvement of the community in the planning process and decision-making. The concept is to involve those people who are most likely to be affected by the decision. Instead of getting a public reaction after the plan has been finalised and approved, it is better to have feedback from the community during the plan preparation process and, if necessary, amend proposals according to feedback received. The downside of public participation is that it can be very time consuming and, if not handled properly, it could be manipulated by specific groups to further their own narrow interests to the detriment of the community.

Urban planning systems are reliant on two main pillars namely forward planning and development control. In forward planning, policy documents and plans are prepared following extensive research and public consultation. The documents give guidelines on how developments are to be carried out. Normally plans are prepared by urban planners within city authorities and then endorsed by the elected politicians.

Development control is that part of the planning process in which applications for development are processed and decided upon. In broad terms, 'development' is defined as the carrying out of building, engineering or other operations for the construction, demolition or alterations on land or sea. Apart from buildings and structures, change of use (e.g. from a garage to a small shop) is also considered to be development even if no alterations are made to the building. National legislations across Europe make it a legal requirement for any development to be covered by a permit. Development applications are assessed with reference to the plans and policy documents that are drafted and approved in forward planning. Successful applications are issued with a development permit that entitles the developer to proceed with the development.

## 2.5 URBAN REGENERATION

Urban areas are complex and dynamic systems. They provide the physical context within which people go about their daily lives. It is where people live, work, recreate themselves and engage in all sorts of activities. Inevitably urban areas are subject to economic, social, physical or environmental problems. Urban problems may be minor or major and may concern just one part of the city or the city as a whole. Urban issues include various combinations of environmental problems (derelict land, redundant industrial buildings, pollution and contaminated land), social problems (poverty, lack of social cohesion, anti-social behaviour, poor housing, poor public facilities) and economic problems (stagnant economy, high long-term unemployment, lack of entrepreneurship).[60]

For the resolution of urban problems, a comprehensive and integrated vision is required supported by action. Urban policy is designed to address urban problems.[61] It sets out a course of action for city authorities to adopt and pursue for the purpose of managing, guiding and bringing about change. Urban policy impacts the physical environment as well as the economic and social conditions of communities. In devising urban policy, urban regeneration is one of the options that is available to city authorities. The word 'regeneration' has become a very widely used term and is used to refer to a variety of processes. Regeneration is very often a main feature of a city's urban policy so it is sometimes difficult to separate the two. The distinctive characteristic of regeneration is that it involves attempts to reverse the decline of parts or all the city.[62]

Urban regeneration strategies were largely developed initially in response to post-war decline of cities and the rising inequality, poverty and unemployment in some parts of cities. In the UK during the 1960s, urban renewal was public sector driven and primarily concerned with large-scale redevelopment of inner-city depressed residential areas. The industrialisation process and subsequent global economic restructuring in the late 1970s and 1980s acted as a catalyst for the development of urban regeneration strategies for many cities in Western Europe and the United States. Regeneration in the 1980s focused on property development using public funds to lever private investment towards large-scale commercially oriented projects, as exemplified by London Docklands.[63] A more recent approach to urban regeneration is to combine in partnership the efforts and the financial input of the private and public sectors to create large-scale mixed developments, possibly with a more heightened environmental awareness than before.[64] This is PPP already referred to in a previous section of this chapter.

In the second half of the twentieth century, urban governance was based on planning systems that provided a policy framework within which urban development, normally by the private sector, can take place. The approach to development was largely seen to be reactive. Proactive action by public authorities was limited to identifying development opportunities which would then be for the private sector to pursue within established parameters set in development briefs or local plans. This is discussed in section 2.3. The urgent need to be competitive induced many cities to take a more proactive approach and adopt an entrepreneurial approach to the urban development within the city. This approach has two strands which in many cities are closely interrelated and overlapping. The first is place marketing whereby cities align the promotion of the city with specific city improvements like, for example, an iconic project, the enhancement of urban spaces or the staging of a major event. This is discussed in section 2.2. The other strand is urban regeneration which is the subject of this section.

A holistic approach to regeneration is reflected in contemporary definitions which describe urban regeneration as 'a comprehensive and integrated vision and action which leads to the resolution of urban problems and which seeks to bring about a lasting improvement in the economic, physical, social and environmental conditions of an area that has been subject to change'.[65] The term 'regeneration' is somewhat vague. It is often viewed as being synonymous with 'revitalisation', 'revival' or 'renaissance'. These terms are often used interchangeably but 'regeneration' remains the most widely recognised and used term by professionals and academics alike. All of the terms have similar meanings and connotations relating to rebirth, bringing new life, revival and reconstitution.[66] Urban regeneration implies that the city as a whole, or at least part of it, was subject to a prior process of degeneration or decay and therefore there is the need to reverse the decline. Since the latter part of the twentieth century, the concept of regeneration has become something of a buzzword in the field of urban planning. Its meaning is usually associated with the redevelopment of cities or areas of cities that have declined physically and economically.

The reasons for embarking on large-scale regeneration projects are varied. At a national level, the government may wish to encourage the economic or cultural redevelopment of former industrial cities that have fallen into decline. Cities view a regeneration project as a means for initiating wider environment improvements and infrastructure improvements. Another motive could be the hope there would be a cumulative effect with the initial funding being a catalyst for further inward investment and the development of other initiatives.

Urban regeneration seeks to bring about a lasting improvement in the economic, physical, social and environmental conditions of an area that has been subject to change.[67] It is a process that aims to revitalise areas of cities that have declined with the specific intent that the benefits are derived not only to the area itself, in the shape of new development and facilities, but also to the city as a whole in terms of increased economic activity and the creation of new jobs.

Approaches to regeneration have evolved over time. Over a period of sixty years, three approaches or agendas have become apparent. Although diverse, these are interrelated and very often overlapping. At different times and in different countries, one of the agendas is predominant depending on the political ideology of the political party governing the city and also depending on the economic and social circumstances at that time. The regeneration agendas are as follows:

- The urban renewal agenda. This focuses on the enhancement of physical and environmental conditions of urban areas, especially those characterised by physical dereliction.
- The social inclusion agenda. This is mostly concerned about the conditions of disadvantaged neighbourhoods. It encourages the development of social cohesion and community participation to bring about the regeneration of communities.
- The economic competiveness agenda. This is concerned with improving economic performance for the purpose of generating employment. The primary focus is on innovation and the increase of productivity and output.

In recent decades, there have been a huge variety of urban regeneration policies, projects and initiatives across Europe. Despite the variety of approaches, Hall and Barrett[68] argue that they have all aimed to achieve one or more of the following goals:

(i) Improvements to the physical environment
(ii) Improvements to the quality of life
(iii) Improvements to the social welfare
(iv) Enhancement of the economic prospects and job opportunities

The overriding objective of city authorities is that city residents would be the main beneficiaries of urban regeneration projects and initiatives.

Urban regeneration projects in cities often involve a cluster of buildings and urban spaces. These may be new buildings on empty sites, redevelopment of existing buildings or the conservation and refurbishment of selected historic buildings. It is not uncommon for an urban regeneration project to include all three. In a study on regeneration, Gospodini[69] analyses cluster-led regeneration projects in inner-city areas using various parameters such as location, economic activities, land uses and architectural morphology. Urban regeneration projects create new 'epicentres' for the city, which Gospodini classifies into four main types:

(i) Entrepreneurial epicentres: clusters of financial services and information technology firms
(ii) High-culture epicentres: clusters of museums, art galleries, theatres, concert halls and other forms of cultural facilities. Several cities have areas that are referred to as 'museum quarters' (e.g. Berlin, Amsterdam and Vienna)
(iii) Popular leisure epicentres: clusters of cafés, bars, restaurants and clubs for music and night-time entertainment (e.g. Temple Bar, Dublin)
(iv) Culture and leisure waterfront epicentres: clusters of culture and leisure activities along waterfronts. The area includes culture facilities, as well as facilities that are solely intended for leisure such as theme parks and promenades (e.g. the South Bank and Bankside, London)

Of the four types of epicentres, three are relevant to tourism as facilities and activities related to culture and leisure enhance a city's attractiveness to tourists.

Gospodini[70] observes that the long-term viability of an epicentre seems to be related to its degree of multifunctionality, that is, to the diversity of uses and economic activities within it. Mono-functional clusters and epicentres are more susceptible to economic shocks and are therefore at risk of rapid decline, even if they are initially successful. In this respect, they tend to be less sustainable than multifunctional clusters and epicentres.

A study on regeneration in European cities[71] analyses three successful urban regeneration schemes in Europe. The report looks at Norra Älvstranden in (Gothenburg, Sweden), Kop van Zuid (Rotterdam, the Netherlands) and Roubaix (Lille, France).

There are several important lessons to be learnt from the success of these three case studies. The first important lesson refers to the need of cities to understand that they are in competition. In each case, a powerful and committed local authority was in charge of the regeneration scheme. The scheme was used not only to improve a run-down or derelict area but, more crucially, to change the image of the whole city and to restructure its economic activity.

The three cities took a long-term strategic approach. They realised that the future prosperity of their city depended upon finding a new economic role. Indeed the regeneration of a deprived area on its own was unlikely to be sustainable. It needed to be linked to the opportunities which wider economic development created. This required clear strategic thinking as well as single-minded commitment. The three cities realised that the world had changed and would continue to do so, because of changing technology and the globalisation of the international economy. These were the forces that had swept away their old industries and that would shape their future.

A second important lesson is that improving the prospect of a city depends on creating places where people would want to live. The case study schemes succeeded in turning what had been low-status areas into attractive ones by improving the quality of their housing. They did this through a combination of refurbishment and high-quality new buildings. This was not just a matter of imaginative housing design but also the way both the blocks of housing and the surrounding areas were planned and managed to create and retain balanced neighbourhoods.

Another lesson is that changing the image is a crucial step in regenerating a run-down area or town. Each of the case studies showed this may take a great deal of effort but regeneration is unlikely to be successful without it. Furthermore, it was important to start making the required investment early on in order to attract the private investment and the new residents that were needed to take the strategy forward. Investor confidence was essential for development and public investment in high-quality infrastructure and the public realm helped to generate confidence.

A fourth lesson is the importance of investment in public transport infrastructure. This is needed not only to generate investor confidence but also to link areas of opportunity in the city with disadvantaged areas. In this manner, the benefits of increased economic activity resulting from regeneration will be better spread throughout the city. Moreover, good public transport facilitates the movement of visitors to the city and thus makes the city more attractive for leisure and other visitors.

Another lesson refers to the role of culture in regeneration. In the three case studies, the city authorities gave particular attention to making their cities more liveable by promoting cultural activities (e.g. those relating to the arts, heritage or local community interests). Culture is important on two levels. First, major cultural facilities and events can play a significant part in transforming the image of an area or an entire city, which is vital if potential investors', employers' and residents' attitudes are to be changed. Second, culture can have other important impacts too, especially if it is used to engage local people, increase their self-esteem and provide access to wider opportunities, including jobs.

The study also derives lessons on the governance that is needed to guide regeneration. The study emphasises the need for a permanent city-owned delivery organisation or development agency to guide the regeneration process. This agency requires a range of knowledge and skills and has to gradually build up experience to be able to navigate through the environmental, social and financial constraints as well as the legal frameworks on planning, land ownership and property development. The regeneration process should focus on the city region rather than just the city. Collaboration is essential between different tiers of government, between neighbouring authorities and between the many agencies and interests involved in regeneration.

When it comes to tourism, cities are in constant need of reinventing themselves, possibly more now following the COVID-19 pandemic. Such reinvention can take the shape of revised urban policies or new regeneration projects or both. One interesting example are the plans for Las Ramblas and the surrounding area which is the oldest and most-visited part of Barcelona. The absence of tourists during the pandemic has revealed Las Ramblas for what it is, a hub with significant cultural potential being the home to three theatres and an international opera house, an art and photography museum and several music venues. The flagship of this cultural transformation will be the grand Teatre Principal, which opened in 1603 but has been closed since 2006. A consortium has raised the necessary funds to refurbish the theatre into a multi-function performance space hosting hi-tech immersive experiences, concerts and other events. The city authorities are eager to bringing local people back to the city's most emblematic street and transform the area into a cultural hub in the city.[72]

## Case Study CC1: The regeneration of Lisbon's city centre

Lisbon is Portugal's vibrant capital. The centre of Lisbon[73] has several historical neighbourhoods where visitors can appreciate the city's built heritage. The city's attractiveness stems from a strong identity based on a rather unique blend of European and Mediterranean culture. Its tourist offer is still mainly centred on conventional city landmarks but this has been successfully extended to other areas of the city, thus offering a broader range of remarkable heritage and landscape sites. According to Santos[74] what makes Lisbon particularly attractive to visitors is 'the diversified character of its urban fabric – from its sensual relationship with topography to its dialogue with the Tagus river'. Added to that is the immaterial cultural heritage such as the Fado and popular festivities.

Over a period of decades, Lisbon has seen a lot of change in terms of urban, geographical and cultural development. Much of that change came about as a result of decisions and investments made by the city authorities with the specific intent to regenerate the city. Lisbon hosted three major events in a space of a decade namely Lisbon 94 European City of Culture, the 1998 World Exposition and the 2004 UEFA European Championship. Coupled with these events were various investments in the city's infrastructure, its urban spaces and also the restoration of its urban heritage. The public transport network was radically improved with the extension of the underground system in the late 1990s. Moreover intermodal stations were established enabling travelers to transfer easily from one mode of transport to another. The overall effect was to shorten the travel times for those wishing to commute from one part of the city to the other. Along with the major restructuring and expansion of the underground network in the late 1990s came the opportunity to redevelop major plazas and squares. The areas close to the new stations were enhanced by means of urban renewal projects thus creating a more cohesive urban structure.

Whether existing or newly extended, pedestrian urban spaces were provided with multi-storey car parking facilities, some of which were located under large squares. Many of the public space improvements were accompanied by the development of food and drink establishments and fashion shops. This displayed an orientation of the improvement projects towards visitors, be they tourists or residents of other parts of the city.

Lisbon's initiatives were motivated by the interests of city residents as well as of visitors to the city. They had a beneficial impact on the daily lives of residents and, at the same time, made the city more attractive for tourism and for leisure-related investments. Moreover tourism and foreign investment have generated renewed interest in the city's urban fabric giving rise to much needed renovations of historic and other buildings.

Across central Lisbon, public spaces and other urban amenities were upgraded and renovated, offering the infrastructure needed to support residential, working and tourist city users. These embellishments were carried out systematically and in a coordinated fashion to eventually establish a continuous and coherent public realm. Santos[75] identifies three networks of embellished public spaces that were established namely the 'riverfront' system, the 'garden and belvedere' system and the 'street and square' system.[76]

Various investments were made to further develop cultural spaces in the city including refurbishment of existing museums and the opening of new ones (such as the Museum of Aljube in 2015). Of particular importance is the Museum of Art, Architecture and Technology (MAAT)[77] located in Belém, at the periphery of the city. The modern design and peculiar curvilinear forms of the building, coupled with its riverside location, gives the building an iconic status – though it remains to be seen whether the building will eventually receive international recognition for its iconicity. Completed in 2016, the new cultural facility has attracted significant visitor flows including many international tourists.

To animate the city and to promote cultural democratization, there have been a steady growth in events and festivals across Lisbon.[78] Although mainly intended for residents, events are also greatly appreciated by tourists – especially if they provide an insight into local culture. An events strategy requires careful thought to avoid the pitfalls of oversupply and excessive festivalisation. Moreover, there is the risk of a potential divide between more local, community-based events and a more globalised commercial offer.

The production of nocturnal cultural events is the link between the city's cultural policy and the nightlife that is offered to residents and tourists. The city has developed a vibrant nightlife scene which is a factor that helps make it more popular with tourists and a section of the local population. Nightlife is often a contentious issue in cities and measures are required to mitigate negative externalities.

International visibility combined with attractive local conditions resulted in a steady increase in the number of international visitors. Public space improvements, combined with investments in transport infrastructure, were the key drivers of urban regeneration in central Lisbon.

## Case Study CC2: Regenerating Sheffield

The city of Sheffield[79] in England was renowned for the production of iron and stainless steel products in the earlier part of the twentieth century and before. International competition in iron and steel brought about a decline in these industries in the 1970s and 1980s. The city needed to find new ways how to generate economic activity and create new jobs for its residents. In the 1980s and 1990s, the public perception of the Sheffield's city centre was poor. It was a compound of piecemeal decision making and urban decay making it a disappointment for anyone visiting the city. The pedestrian connection from the train station to the city centre was especially disappointing. Those arriving by train to the city were, upon leaving the station, brought face to face with a scruffy car park. Then they had to cut across Castle Square which was a heavily trafficked roundabout with an unpleasant shop-lined underpass below it. The rest of the walk to the centre was up a street with various semi-derelict properties.

The approach of the city authorities was to create a greatly enhanced strategic pedestrian route connecting the city centre to the train station. This was achieved by upgrading the urban spaces along the route and concurrently redesigning the city centre spaces – most notably the urban space referred to as the Peace Gardens. This was remodelled in the shape of a shallow amphitheater focused on fountains that rise from the floor of the space with one side of the town hall building acting as a backdrop. Another urban space that was redesigned and upgraded was Sheaf Square in front of the train station. The creation of high-quality public realm was financed in part by commercial development that replaced a 1970s public building which was the extension to the town hall. Two visitor attractions were created: the Millennium Galleries (an art gallery and museum) and Sheffield Winter Gardens (see Case Study VA4 in chapter 5). All these various projects were completed within an eight year period between 1998 and 2006. The Peace Gardens, the Winter Gardens, Sheaf Square and the Millennium Galleries are widely considered to be successful as evidenced by the extent they are used by the public and the way in which these spaces appear to have extended the hours during which the city centre is actively used. The care and quality of the design of these spaces is a major factor in their success. Also important is the way these spaces connect with each other.

Another important development in Sheffield was the Supertram, the new light rail that ran its first services in 1994 on a line running from the city centre to Meadowhall shopping centre. The Supertram required the remodeling of a series of public spaces in the city centre and the downgrading of a major road that passes through Castle Square. The underpass at Castle Square was filled in and the ground surface was redesigned to create a space more amenable to pedestrians and to the users of the light rail. With four lines and fifty stations, Supertram carried 15 million passengers in 2017.[80] The successful setting up of a mass transport system was an essential precursor to the urban regeneration that was happening at Sheffield.

The above discussion refers primarily to city centre interventions and public transport. There were other initiatives and investments in Sheffield that were essential elements in the city's regeneration.

The World Student Games were staged in Sheffield in 1991. These were poorly managed and ended up with substantial financial losses. On the plus side, it left a legacy of two major sports facilities namely the Ponds Forge International Sports Complex and the Don Valley International Athletics Stadium. These and other facilities enable Sheffield to host a wide range of national and international sports events. The multi-purpose Sheffield Arena is another legacy of the Games. Apart from hosting indoor sports events, the arena is used for major concerts and large conferences.

In 1990, Meadowhall shopping centre was opened. It occupied a site in the former industrial zone at the periphery of the city – in an area that was previously occupied by heavy industry and steelworks. With a floor area of 140,000 square metres, it was, at the time, the second-largest shopping centre in the UK. The massive increase in retail floor space at the city's periphery set the city centre into a downward spiral with shop closures and less customers. A declining city centre was the catalyst resulting in the investments in the improvement of the public realm that were described above.

Sheffield is home to two universities, University of Sheffield and Sheffield Hallam University. With a combined student population of over sixty thousand, the role of these two universities in the city's economy cannot be understated. They are major employers and they have invested heavily in new buildings and facilities to cater for a constantly growing number of students. After gaining university status in 1992, Sheffield Hallam invested in the consolidation of its campus in an area close to the train station.

*Chapter 3*

# The Changing Dynamics of Tourism and Leisure

Cities are constantly changing due to a range of social and economic forces. These were discussed in the previous chapter. Tourism is a further force of change for cities and the increasing demand for tourism and leisure has made this force all the more influential. Public authorities invest in the public spaces and the infrastructure of a city primarily to cater for the needs of residents and enable it to cope with increases in visitor numbers. Growth in tourism motivates the private sector to invest in facilities that tourists use such as hotels, catering outlets, visitor attractions and so on.

The dynamics of tourist activity have changed in the past three or four decades. They have changed in a manner that increases the effect of tourism activity on cities. This chapter discusses these changes; such a discussion is needed for a better understanding on how tourism changes cities.

## 3.1 CHANGING TOURISM PRACTICES

In recent decades, significant social and economic changes have brought about an increased demand for leisure as well as for more overseas travel. Greater affluence enables many to dedicate a greater part of their income to leisure activities. Changing lifestyles, more vacation leave and more flexible work practices enable people to travel more frequently for leisure and also for other purposes. Whereas the norm used to be one big family holiday in the summer, nowadays it is normal for people to travel away from home twice or thrice each year, or possibly more.

There is the increased tendency of Europeans to take additional but shorter holidays with city breaks becoming increasingly popular. An important contributor to this trend was the increased availability of low-cost air travel with its emphasis on short haul, point-to-point journeys.[1] A wide range of city destinations became more easily accessible at a lower cost. The increased demand for city breaks was also fuelled by people's changing perception of cities as travel destinations. The city is increasingly perceived as a destination in its own right rather than just as an entry to or an exit from the region or country.[2] A city break represents a distinctive type of holiday that can be defined as 'a short leisure trip to one city or town, with no overnight stay at any other destination during the trip'.[3] This definition highlights the 'city only' nature of the trip.

## 3.2 THE CHANGING ROLE OF CULTURE IN TOURISM

The relationship between culture and leisure has evolved over time. Museums, theatres and other cultural facilities were initially supported by the patronage of the elite class. Cultural production happened only because there were wealthy individuals willing to support it for the purpose of building the patron's social status and reputation. This phase of cultural development is referred to by Sacco[4] as 'Culture 1.0'.

As societies evolved, there was an increased willingness of people to pay for some form of cultural entertainment. There was a shift towards 'Culture 2.0' where culture increasingly became an educational and economic field. Public authorities devoted resources to the support of culture and the arts to the benefit of the society as a whole.

The next phase of development, 'Culture 3.0', is brought about primarily by the diversification of cultural taste and access to new technologies and media. Alongside economic value, culture is also seen as a means of creating identity, stimulating social cohesion and supporting creativity.[5,6]

The way people engage with culture and leisure has changed and this has brought about changes in the way people engage with tourism. The changing relationship between culture and leisure was mirrored by an evolution in the role of culture in tourism. In the nineteenth and twentieth centuries, nation-states invested in cultural facilities, such as major museums, in order to support an idea of national identity and cohesion. In the latter part of the twentieth century, these facilities became part of the tourism system as culture no longer remained the prerogative of those with a keen interest in arts and culture. It became of interest also to a wider audience including urban tourists.[7] Tourists' increasing demand for cultural facilities made national governments and city authorities more aware of the economic potential of culture as a means for encouraging tourist spending at the destination. In the 1990s, this led to a boom in investment in cultural facilities.[8] Many cities built new museums as a means of stimulating economic growth, both directly by attracting tourists and indirectly through improving the city's image.

The way people engage with culture and leisure has changed and this has brought about changes in the way urban tourism is practised. Richards and Marques[9] claim that 'cities are seeing a shift from two separate systems of "culture" and "tourism" towards the integrative phenomenon of "cultural tourism", and are increasingly moving towards a "culture of tourism", in which tourism becomes one of the major modes through which increasingly mobile populations interact with the urban environment'.

## 3.3 BLURRING OF THE BOUNDARIES BETWEEN CITY RESIDENTS AND TOURISTS

The traditional view of the city destination is that of a city that is predominantly residential with a few hotspots of intense tourism and leisure activity. Such tourism hotspots are located in areas with a cluster of visitor attractions, entertainment facilities or hotels. Underlying much of the past academic discussions of tourism in cities was an implicit assumption that it makes sense to see visitors as separate from the host community – wanting different things from the city and thus behaving differently. This was the predominant model for urban tourism up until the first few years of this century. This urban tourism model has changed as sustaining this separation is becoming more problematic.

Nientied[10] distinguishes two types of urban tourists. There are those who have a list of must-see highlights, and others who value experience over the main sights. The latter type prefers places in the city that are off the beaten track. They are eager to leave the confined space of the tourist bubble to 'truly' experience the city. Nientied labels this trend as 'new urban tourism'. Local experiences are increasingly sought and consumed by the new urban tourist. This desire distinguishes new urban tourism from the traditional, mass tourism.

In the more well-established city destinations, it is increasingly difficult to identify tourism as a discrete activity contained tidily at specific locations and occurring at set periods. There is an increasing overlap between what tourists do, what day visitors do and what residents and workers do on their evenings and days off. The main duality between residents and tourists disappears. Instead, urban areas are characterised by a vast range of 'urban users', including tourists and thus tourism activity becomes integrated into the urban life.[11]

There are two social and cultural processes that have brought about this change. The first relates to the lifestyle of city residents. Greater affluence and changing lifestyles have enabled more people to dedicate more time and money to travel and also to leisure activities in their home city. When city residents aren't travelling themselves, they engage in similar activities as the tourists: going out to eat, walking the streets and along promenades, visiting museums.[12] In this way, residents behave more and more like tourists in their own city.[13] In terms of leisure activities, the boundaries between tourists and city residents are increasingly blurred.[14] This is illustrated in research carried out on visitors to four attractions in Barcelona. Tourists were more likely than residents to visit individual attractions such as the Antoni Gaudí sites La Sagrada Familia and Parc Güell and residents were slightly more likely to have visited districts of the city such as the Barri Gòtic or Port Olímpic. These figures indicate a considerable spatial overlap between tourists and residents.

The second refers to the way the tourists engage with the city destination. Some visitors are drawn to areas of the city that are not the traditional touristy areas. These are the 'new urban tourist' as described earlier. Their main interest is to explore the city and to seek experiences away from planned tourism areas and well-established tourism zones.[15] Some cities have latched on to the growing desire of tourists to experience the local lifestyle. 'Live like a local' has become the new creed promoted by many cities' promotional websites.[16]

In the past decade, there was another process that brought about the spread of tourism activity into residential areas; short-term rental tourism accommodation. These have grown exponentially following the advent of large platform companies such as Airbnb. The highest densities of short-term rentals are located in city centres and around major attractions although such tourism rentals also extend to residential areas.[17] By staying in a short-term rental the tourist can gain the experience of a local resident.[18] This is in synch with the approach taken by some cities to encourage tourists to get a glimpse of the life of local residents.[19]

As time passes, more and more cities are gradually evolving away from the traditional 'tourism hotspots' model to a city life where the activities of residents and tourists become increasingly indistinguishable in many parts of the city. The blurring of boundaries between residents and tourist activities occurs concurrently with an increasing presence of tourists and other temporary residents in the city. Such a combination changes the way city residents see themselves in the world. Rather than just solely as residents of the city, residents begin to think of themselves more and more as part of a European and an international society.

In the coming years, the dispersal of tourism activity across the city will persist due to the COVID-19 pandemic as some visitors to cities will try to stay away from tourist hotspots that are perceived to be crowded and hence less COVID-safe. They will seek out other parts of the city that are attractive and that provide leisure activities, thus bringing about further mixing of tourists and residents.

## 3.4 IMPROVED CONNECTIVITY BETWEEN CITIES

Socio-economic changes combined with improvements in aviation acted as catalysts for substantial increases in overseas travel. In the late twentieth century, tourism evolved dramatically because of developments in aviation and air travel such as larger aircraft, more air routes and improved services for travelers including charter services and package tours. These increased the travel options available to the tourist and the range of destinations that they can visit. Many areas of the world were opened to tourism, particularly the tourism resorts and cities in South Europe along the Mediterranean coast.[20]

In the past twenty years, changes in the way the aviation industry operates have brought about further changes in tourism destinations worldwide. The introduction of low-cost carriers (LCCs) has made travel cheaper. More crucially, the business model upon which LCCs is based has made it possible for a large number of new air routes to be opened up across Europe. The opening of a new route is determined through negotiations between an LCC and the local authorities. It is common practice across Europe for airport or national authorities to provide incentives to airlines to fly to their airports – the more attractive the incentives, the more likely that the LCC will agree to launch a new route.

Most major cities have important airports which are well connected. Major cities have large populations and act as gateways to the rest of the region. Up until the end of the century, air travel relied almost exclusively on the major airports close to the larger cities. That has changed since the beginning of the millennium with a significant number of secondary airports across Europe handling a much larger number of passengers flying on LCCs.

Cities that are within easy reach of an airport are more accessible because no change in the mode of transport is necessary. A city near an airport is much more accessible than one which is not. Many cities across Europe are now much more accessible thanks to the increased connectivity provided by LCCs. Small- and medium-sized cities now have the opportunity to compete with more popular destinations. With increased tourism numbers worldwide, the secondary city destinations can now partake a larger slice of what has become a larger cake.

Historically the two main airports that serviced Milano and the surrounding region were Milano Linate and Milano Malpensa. In 2000, these managed 6 million and 21 million passengers, respectively. Passenger numbers grew to 9 million and 25 million, respectively, in 2018. In the meantime, a third airport,

Bergamo – located just outside Milano, to the east – came on the scene. Despite having just 1.2 million passengers in 2000, passenger numbers grew rapidly to reach 13 million in 2018, making it the third-largest airport in all Italy.[21,22] The main reason for this growth was that LCCs took advantage of the cheaper landing rights fees that were on offer in this secondary airport. Increased passenger traffic through Bergamo airport enhances the accessibility not only of Milano and Bergamo themselves but also of several towns and cities to the east of Milano such as Brescia, Verona, Trieste and Mantova. Bergamo is a charming town with significant tourism potential. Having an airport on its doorstep enables it to capitalise on its many attractive attributes. The same can be said for other towns and cities in the region, albeit to a lesser degree because some land travel is required.

The LCC business model seeks to lower ticket prices by cutting costs on customer service, ticket distribution and optimal fleet utilisation. The profits are made not on the flight tickets but on the sale of ancillary services; services that are beyond the transportation of the passenger from A to B. These include the in-flight sale of food and beverages as well as priority boarding and seat reservations.[23] The LCC business model relies heavily on high utilisation of the aircraft, meaning that for each flight, the aircraft is full of passengers. This is achieved by offering flight tickets at heavily discounted prices days or weeks before the flight if and when the tickets are not being sold sufficiently. Having flights full of passengers makes for a close correlation between tourists arriving by air to a destination and the number of LCC flight routes that are offered from its airport.

A look at Malta's tourism numbers clearly illustrates this. Although an island, Malta is comparable to a medium-sized city with a population of around half a million. It is an established sun and sea destination in the Mediterranean. Low-cost airlines were introduced to Malta in 2006. At the time, the destination was connected to seventy-nine airports, mostly across Europe and the number of tourists visiting Malta stood at around 1.1 million per year. Those numbers increased to ninety-four airports and 1.8 million tourists in 2015. In 2015, the Maltese government started pursuing an aggressive policy of establishing new air routes by offering attractive incentives to LCCs. The numbers in 2018 were 113 airports with direct connections and 2.6 million tourists.[24] The correlation between the number of air routes and number of tourists is evident.

Improvements in land transport are also crucial to make cities more accessible to potential visitors. Compared to two or three decades ago, it is much easier nowadays for people to travel from one city to another in Europe. There have been massive rail and road investments across Europe, very often supported with substantial funds from the European Union by means of the Trans-European Transport Network (TEN-T) policy. TEN-T aims for the development of a Europe-wide network of transport infrastructure based mainly on rail and road but also on inland waterways, maritime shipping routes, ports and airports. The declared objective is 'to strengthen social, economic and territorial cohesion in the EU'.[25] By implication, this objective seeks to make cities across Europe more accessible to visitors and therefore more amenable to tourism activity.

The Commission has established ten 'corridors' stretching across the EU along which significant funds have been invested – and will continue to be invested – on transport infrastructure projects. For example, the Baltic-Adriatic Corridor is an important trans-European road and railway axis connecting the Baltic seaports of Gdańsk, Gdynia, Szczecin and Świnoujście in the north with the Adriatic ports of Koper, Trieste, Venice and Ravenna in the south. Along the way the corridor includes several important cities in several countries. The cities that will derive economic and tourism benefits from improved transport infrastructure along this corridor include Warsaw, Poznań, Katowice and Wroclaw (in Poland); Bratislava (Slovakia), Vienna and Graz (Austria); Maribor and Ljubljana (Slovenia); and Udine, Padova and Bologna (Italy). A network of corridors, such as the Baltic-Adriatic, enables access to the four corners of the EU.

Travel has also been made easier with new technologies that allow online booking and payment. An improved road infrastructure facilitates passenger transport services by means of coaches. Inter-city services have increased. They provide a cheaper option to rail and better access to towns and cities that might not be so well serviced with rail. Improved transport infrastructure also signifies greater reliability including adherence to time schedules and reduced delays. Because of the freedom of movement between EU states, there are no border bureaucracies to deal with. Greater ease of travel facilitates tourism and thus enables more and more cities to derive benefits from tourism.

### Case Study CD1: The reconstruction of Rotterdam Centraal

Rotterdam Centraal railway station[26] is one of the most important transport hubs in the Netherlands. The 1957 station consisted of a stand-alone building at the front with a tunnel and open platforms, each with their own canopy. The former station was no longer able to handle the estimated one hundred and ten thousand daily rail passengers, nor the complexity of a major transport hub. In 2004, the Municipality of Rotterdam initiated the process for the total reconstruction of the station and its surroundings. The old station building was demolished in 2008 and the new train station became fully operational in 2013. The new building has all train platforms and station facilities under one roof, creating a single entity connecting the platforms to the main entrance hall and passenger thoroughfare.[27] The roof above the platforms is made from glass so passengers arrive into an airy space filled with natural daylight. The new station is larger, brighter and provides for passenger flows that are more orderly and organised than previously. The number of daily travelers at Rotterdam Centraal is expected to increase to approximately over three hundred thousand daily by 2025.

Rail services connect Rotterdam to cities across the Netherlands and also provides international services to London, Brussels, Paris and Marseille. Moreover, the station is inextricably intertwined with the city's urban transportation network, making it easy for passengers arriving at the station to travel to any part of the city using underground, light rail or bus. The esplanade in front of the station, Stationsplein, is a continuous public space free of any traffic. A parking garage for seven hundred and fifty cars and a bicycle shed for more than five thousand bicycles are located under the square. The pedestrian esplanade extends over a hundred metres in the direction towards the city centre, connecting with other pedestrian friendly spaces of the city.

In terms of architectural design, one of the main challenges of Rotterdam Centraal station was the difference in the urban character of the north and south side of the station. A modest design was created for the north entrance appropriate for the character of the neighbourhood. In contrast, the grand entrance on the city side is clearly the gateway to the high-rise urban centre. Here the station derives its new international, metropolitan identity from the hall made of glass and wood. The roof of the hall, clad with stainless steel, gives rise to building's iconic character.

To address sustainability, the roof is fitted with over one hundred and thirty thousand solar panels covering about a third of the total twenty-eight thousand square metres of roof area. The solar cells have a high degree of transparency thus allowing ample light into the building and reducing the need for artificial lighting. The cells generate three hundred megawatts per annum, which is equivalent to the energy consumed by about a hundred households. The use of solar panels reduces the station's carbon dioxide emissions by about 8 per cent.

Transport has been an important element in Rotterdam's renaissance.[28] Apart from this major investment in Rotterdam Centraal and the surrounding area, the city also invested in a high quality tram system and built the Erasmus Bridge over the Nieuwe Maas which is a distributary of the Rhine River (figure 3.1).

## Case Study CD2: Billund Airport, Denmark, and the effects of low-cost airlines

Billund is located in central Jutland in Denmark. The town and municipality have a population of twenty-six thousand. It is the home base of the Lego Group, an international company whose flagship product, Lego, consists of variously coloured interlocking plastic bricks. Legoland Billund is one of

**Figure 3.1    Rotterdam Centraal.** *Source:* Author's own.

several amusement parks that have been developed internationally under the Lego brand.

Billund Airport started its operations in the 1960s upon the initiative and investment of the Christensen family, owners of Lego. The airport is a short distance away from the town of Billund and less than an hour's drive from various towns and cities in central Jutland including Aarhus, Horsens, Vejle, Kolding, Esbjerg and Fredericia.

The manufacturing sector in Jutland was in decline since jobs were lost as the industrial production of many products was outsourced to less-developed countries with lower wages.[29] To counter the job losses, Billund municipal council opted for an economic strategy based on tourism and the experience economy. The strategy resulted in several capital projects in the area surrounding the airport, including the improvement and expansion of Legoland and additional facilities to the airport itself. Passenger numbers handled by Billund airport increased from seventy thousand in 2006 to just under two hundred thousand in 2008.[30] This sharp increase was due to the rapid growth of low cost carriers (LCCs) at the time across Europe.

More crucially the Billund strategy resulted in the development of Lalandia Billund, a water park and holiday resort. Completed in 2009, it occupies an area of four hectares close to the airport and also close to Legoland. Hundreds of jobs were created not only at the entertainment facilities but also in the various support ancillary facilities and services that tourists make use of during their visits. A synergy was developed between the airport – with the ease of access that LCCs provide – and the provision of extensive entertainment facilities in close proximity to the airport. The development of the Billund airport due to LCCs has also been beneficial to central Jutland as towns and cities in the area can now consider tourism growth as a viable option.

Apart from domestic flights in Denmark, there are direct international flights from Billund to at least forty-two cities in twenty-six countries[31] as well as several Mediterranean seaside resorts. Passenger numbers continued to grow steadily from 2.8 million in 2012 up to 3.8 million in 2019.[32] The entertainment sector in Billund became well established with almost 2 million visitors to Legoland Billund in 2019 making it the largest tourist attraction in Denmark outside Copenhagen. In the same year, seven hundred thousand people visited Lalandia putting it amongst the top eight water parks in Europe.[33]

*Chapter 4*

# City Resources for Tourism (1)

## *Areas in the City*

#### 4.1 THE TOURISM PRODUCT

The range and quality of the resources offered by a destination is one important factor that attracts tourists to it. A tourist at a city destination makes use of a range of resources and therefore the overall experience of the destination is the outcome of a combination of attractions, transport, accommodation, entertainment and other services. Whether bought as a package from the travel agent or assembled by the tourists themselves, the tourism product is in practice an amalgam of many different products and services supplied by different entities. The fragmented nature of supply creates the challenge for a destination to achieve coordination and integration of all components across all sub-sectors of the tourism industry.[1]

A tourism resource is any facility which is, or could be, used by tourists.[2] Combined with other elements, these can also be collectively referred to as tourism product. Tourism product can be defined in different ways. One definition is that the product 'embraces all elements with which the visitor to a destination comes into contact, including infrastructure (for example transport and utilities), the service personnel, places of lodging, attractions and activities, facilities and amenities'.[3] Another possible definition is that the tourism product refers to those attractions, activities and facilities that are specifically provided for the visitor. The former definition is preferable as there are many facilities and features in a city that were not developed or created specifically for the tourist but which impinge on the tourist experience and are therefore part of the tourism product. Moreover, without an adequate infrastructure, a destination's tourism product will fail to meet the expectations of the tourist. The infrastructure is essential to support the development of the things tourists want to see and do on their visit.

The tourism product can be classified into the core product (primary elements) and the supporting facilities and services (secondary elements).[4,5,6] The core product is that whose benefits motivate tourists to visit the city. This includes historic buildings, urban landscapes, museums, art galleries, concerts, spectator sports, conferences, exhibitions, special events and unique shopping opportunities. On the core product, a further distinction can be made between activity places and leisure settings. The former refers to places where tourists and residents can engage in some kind of leisure activity. The latter is discussed further on this chapter.

The supporting products (secondary elements) are the facilities and services that tourists consume and that provide support to the tourist visit. They are the additional services that enhance the visitor experience but are not enough on their own to motivate a visit.[7] They include accommodation, catering, shopping and transport. They can be critical for a city destination to be successful but very rarely are they the reason for the visit. Few people would travel long distances because a city has good hotels, a varied range of restaurants, interesting shops or an excellent transport infrastructure. However, they might be deterred from visiting a city if there is a shortage of reasonably priced hotel rooms or if it is exceedingly difficult to move around the city. Although not the reason for the visit, supporting services impinge on the overall tourist experience of the city destination. Knowing that these elements are easily available and offer value for money may provide an additional reason for visiting the city or tip the balance in the choice between one city and another. Thus, the provision of these elements is essential for the development of tourism in a city.[8] Supporting products require careful consideration because they involve a significant part of the tourism expenditure. In Britain, in 1998, the expenditure on accommodation, food and drink, shopping and in-UK travel accounted for as much as 91 per cent of total tourist expenditure.[9] Effective provision of supporting products will maximise the economic benefits that will result from tourism.

The aforementioned classification is useful but it is difficult to fit all product elements neatly into these categories.[10] In some cities, the so-called secondary elements of shops and restaurants may well be the main attractions for certain groups of visitors.[11]

The distinction between core product and supporting services does not imply a more intense or more exclusive tourism use of the former. On the contrary, tourists may visit a city for its primary attraction but spend most of their time and money on supporting facilities.[12] Nor does it imply that the core products were created for tourists while supporting facilities were not. St. John's Co-Cathedral in Valletta and the Wawel Royal Castle in Kraków were not created for tourism but are primary attractions to many. On the other hand, tourists are the intended clientele of most hotels and many restaurants, cafes and souvenir shops. Moreover, there are many elements of the city for which the tourist use is only a very small part of their overall demand.[13] In many instances, they are used predominantly by city residents (as places of worship, transport facilities, offices, commercial use, etc.), while at the same time they are part of the tourist itinerary for the city. An example of this is a heritage building that has been converted to office use. Tourists may admire the building from the outside and get to know narratives about the building. Such tourist use is marginal compared to the building's commercial function. Tourists make use of a wide variety of facilities but an exclusive use of very few. Equally these facilities are rarely produced for tourists but are shared by many different types of users; 'in short the multifunctional city serves the multimotivated user'.[14]

It is normal that the demand for a destination fluctuates over the course of a year. Tourism operators have periods when demand is near capacity and others where the level of utilisation is below their break-even point. This phenomenon, termed as seasonality, is most often associated with beach and ski resorts but it is also present in city destinations, albeit less pronounced. Demand variations are problematic primarily because a service cannot be stored. For example, if a paid visitor attraction can handle 10,000 people per day, then having just a thousand visitors on a particular day will mean that the opportunity for generating income on that day has been utilised only up to 10 per cent. The remaining 90 per cent is lost and cannot be retrieved after that day has passed. The same can be said about flights, hotels, restaurants, conference facilities and so on.

Reduced tourism demand over part of a year signifies an underutilisation of the tourism facilities. This is the reason why destinations take initiatives to boost tourism demand during the off-peak. In particular, destinations seek to create attractions and activities that are specific to off-peak periods. This could include staging of events (e.g. culture-oriented festivals or sports events) or the construction or revitalisation of tourism facilities (e.g. conference venues, theatres, museums and visitor attractions).

At any point in time, a city's tourism product is the outcome of a multitude of decisions taken by public authorities and by private investors over previous decades. City authorities have a particular important role in product development within their overall objectives of encouraging economic growth and facilitating those sectors that can have a competitive advantage over city competitors. For most tourism facilities and resources, authorities are not directly involved in the development and management. The authorities' role is to create the right financial and social conditions for the private sector to invest and develop commercial facilities and services, for tourists as well as locals.

Tourism by its very nature cuts across many parts of city governments. City authorities normally have departments to deal with specific responsibilities,

Urban planning, culture, heritage conservation, capital projects, transport infrastructure and environmental protection are essential responsibilities for the purpose of improving the quality of life of city residents but they are also relevant to tourism particularly in terms of their impact on the tourism product. These departments within city authorities normally have significant financial and human resources at their disposal. Even if not specifically intended for tourism, the manner of use of these resources will have a major impact on the way the city's tourism product evolves over time. The complexity of the cross-cutting nature of tourism gives rise to the problem of coordination especially on matters of tourism product development.

Some city authorities have departments or agencies with a specific responsibility for tourism. These are referred to as destination management organisation (DMO), even if in most instances, their primary role is marketing. Apart from marketing, a DMO engages in market research, licensing of tourism/leisure establishments, tourism information and sometimes small-scale improvement projects. A DMO seeks to take a more proactive role in fostering and managing the benefits of tourism development.[15] Another important role of a DMO is to lobby the big-budget departments within the city authority to take tourism into consideration in its plans and investment decisions. The objective is for major investment decisions to be taken in line with the city authority's own tourism plan for the city. For example, when the city authority is preparing an urban plan for part of the city, the city's DMO should seek to give its input. Similarly, if the transport department is planning a new investment in public transport infrastructure, the DMO should seek to negotiate changes that will facilitate the movement of tourists to and from major visitor attractions.

In a city, there are many companies and small businesses operating in tourism, hospitality and leisure. These are the backbone of a city's tourism product because they produce the services that tourists consume at the destination. Characteristically their products and services target principally the tourist markets. Moreover, they have a high level of interaction with tourists. Collectively, however, these enterprises do not produce and deliver all of what a tourist consumes while visiting a destination because the overall experience is an amalgam of the various services. Much of the quality of the visitor experience stems from non-commercial activity such as enjoying the climate and scenery, mixing socially with local inhabitants, using free public services and infrastructure and strolling in public spaces.[16]

Because of the multiplicity of interested parties, successful tourism planning requires a balance between the various needs and values of the entities involved and building trust between them. The process of consultation and consensus building is thus crucial for the development of tourism and the attractions, activities and facilities that are provided for visitors. Public consultation can take many forms: surveys, public meetings, group discussion with stakeholders, media features and feedback. To be effective such consultation needs to take place at, and between, all levels of the administration, with the private sector and with the local communities where the tourism activity takes place. Coordination can only be achieved if all parties recognise the mutual advantage of working together and joining forces in designing, delivering and marketing tourism products. Coordination between neighbouring towns and cities is also advisable to promote tourism activity within the region – tourism from which the towns and cities stand to benefit.

In cities whose tourism sector is not yet well-developed residents are likely to be eager to see the tourism sector grow to create new job opportunities and also to reinforce their pride in the city. Public consultation is likely therefore to generate a lot of interest provided that it is guided by city officials in a genuine manner to listen to the residents' desires and concerns. Residents in cities with more developed tourism activity are likely to take a different stance with concerns being expressed on how the tourism activity is impacting them because of overcrowding at key locations, impacts on housing market and so on. Caution is needed to ensure that the consultation process is not dominated by one particular lobby group. This could be a vociferous environmental non-governmental organisation (NGO) or commercial interest groups with strong political connections. In the latter case, commercial interest in real estate development could dominate the city's agenda to the detriment of the promotion of sustainable tourism activity in the city.

A publication by the UN World Tourism Organization and the European Travel Commission (*Handbook on Tourism Product Development*)[17] recommends to public authorities to give greater priority to tourism product development (TPD). It notes that many public authorities prioritise tourism marketing while giving much less importance to TPD. This comes about partly because tourism operators often pressurise governments and city authorities to increase financing for tourism promotion. Increased tourism numbers will inevitably result in more occupancy in hotels and greater usage of other commercial tourism facilities.

The handbook gives reasons why TPD should be given higher priority. In many cities, tourism is one of several sectors that contributes to a city's economic growth. TPD is also a means of minimising damaging environmental and socio-economic impact on the destination and on the resident community. A third reason is that TPD improves the provision of facilities that the local population can enjoy, thereby improving their quality of life. As the living standards of city residents improve so does demand for leisure and recreational provision. TPD can meet this need while at the same time providing facilities that will make the city destination more attractive for visitors.

In many instances, local demand alone is not enough to financially sustain a leisure facility and it would be only thanks to outside visitors that the facility can run at a profit. Examples of this include theatres, theme parks, transport services, events and concerts.

Operators in the tourism and leisure sectors often pressurise national governments and city authorities to develop the enabling transport and utility infrastructure and to establish vocational training systems. This is especially so in countries that are in the development phase of tourism. Operators often encourage authorities to create a positive image of the destination and to offer incentives to small- and medium-sized enterprises as a means of developing the supporting ancillary facilities for tourism to grow.[18] Operators often argue that public authorities have the responsibility to finance the marketing of the destination as a whole.

For any product development, the attractions, facilities and events must first and foremost be in line with the tastes and trends in the market. Destinations need to know what the market likes and what it is looking for as well as how its preferences are changing and are likely to change in coming years. Demand is not static and is subject to a wide range of influences, including marketing and promotional campaigns. The challenge for destinations is to translate these patterns of demand into products and their presentation. This process involves relating the resources available in the destination to the market demand and developing a tourism product offer that will appeal to a number of identified markets and segments.[19]

Cities that are in the process of developing their tourism sector require a system for the collection, analysis and interpretation of tourist statistics related to the destination. The data can be gathered from registered or licensed tourism accommodation, attractions and leisure operations. Carefully planned surveys amongst visiting tourists are another means for collecting useful information on the tourism activity within the city. Moreover, information on prospective source markets can be derived by means of regular market research. There are different market segments with different needs and expectations. A destination needs to distinguish between different demands from short-haul and long-haul markets, from first time and repeat visitors, from leisure tourists and business persons, etc.[20]

A large city has an advantage in developing the tourism sector.[21] People are already traveling to large cities because they are regional, national or even international centres of business. They already have accommodation and catering establishments for visitors. In addition, they have a transportation system that gets visitors around the city. It is relatively easy for these cities to simply add capacity to also handle leisure tourists. Providing tourist services can be a challenge for small cities and towns as they may need to establish transportation, accommodation and food services for visitors as they do not already exist. Or they may exist but the quality and choice available is insufficient to support a thriving tourism industry. The same can be said for cities that are close to popular beach resorts in the Mediterranean. To support the sun and sea tourism, accommodation, tours and other tourist-oriented services would be well developed. It is not uncommon for cities along the Mediterranean coast to seek to diversify to culture-oriented tourism, using their sun and sea tourism infrastructure as a basis.

Specht[22] focuses on the more tangible aspect of the tourism product and advocates a segmentation based on intention, type and function of architecture. In an adaptation of Specht's model (figure 4.1) facilities and resources are included with architecture to provide a more holistic representation of the tourism product. In terms of intentions the architecture/facilities/resources may be developed to meet the demand of visitors only or of visitors and locals. In terms of types, building and architecture within the city can provide for a wider range of facilities and resources, as listed in the second row of the table. The third row gives examples.

Depending on its characteristic, the same architectural structure might serve a variety of functions. For example, accommodation and amenities are often combined whereby hotels often host restaurants and shops. In some instances, large casinos and theme parks form part of a larger integrated functional ensemble. Hence, primarily representing leisure and recreational facilities, such developments may eventually cover a broad range of functions

| Intention | Architecture/ Facilities/ Resources developed or made available for demands of visitors | | Architecture/ Facilities/ Resources developed or made available for demands of locals and visitors | | | Architecture/ Facilities/ Resources developed or made available for demands of locals |
|---|---|---|---|---|---|---|
| Type | Accommodation | Infrastructure (for access and mobility) | Amenities | Leisure and recreation facilities | Cultural facilities | Other facilities |
| Function (example) | Hotel | Footpaths and walkable streets | Coffee shops | Public open space | Museums | Educational buildings |
| | Holiday apartments | Public transport – buses | Bars | Green parks | Urban heritage | Industrial buildings |
| | Shared accommodation | Public transport - Light rail | Restaurants | Theme parks | Art galleries | Religious buildings |
| | Boutique hotel | | Banking services | Sports venues | Theatres | Offices |
| | Bread & breakfast | Road networks | Shops and department stores | Zoos | Concert halls | |
| | Hostel | Train Stations | | Aquarium | Monuments | |
| | Camping site | Railway Infrastructure | | Casino | Exhibition spaces | |
| | | Airport | | Night clubs | | |

Figure 4.1 Functional segmentation of city facilities and resources. *Source:* Adapted from Specht, 2014.

from accommodation to transport infrastructure to amenities to cultural facilities. 'Other facilities' such as offices could form part of these facilities. As a part of transport infrastructure, airports and train stations often host various amenities, sometimes including accommodation. An unusual combination is that of a bridge that functions as a retail area as is the case with Ponte Vecchio in Florence. The medieval bridge still has shops along it, as was once common. Occupied initially by butchers, today's tenants are jewellers, art dealers and souvenir shops. At the same time, the bridge is one of Florence's main tourism attractions.[23]

The resources and contexts referred to previously are often talked about in tourist and promotional websites of city destinations. Known attractions are included in the suggested itinerary for visitors or at the very least get a mention in promotional material for tourists. It is not uncommon, however, for visitors to a city to encounter city features and places that are not included as attractions in the promotional material but which nevertheless are perceived charming by the visitor and are enjoyed by them.[24] One example that comes to mind is a residential area that may have interesting architectural or natural features. The relevance of these unspoken features and places should not be underestimated. These too are part of the tourism product even if they are not widely considered to be so by tourist stakeholders.

## 4.2 TOURISM DISTRICTS AND LEISURE SETTINGS

Towns and cities offer a wide range of visitor attractions and tourist facilities that tend to be spatially concentrated. A common occurrence is the transformation of one or more urban areas because of tourism activity. Such areas generally develop a distinctive character due to a particular mix of activities and land uses. These could be restaurants, hotels, visitor attractions, evening entertainments or shopping streets. A concentration of tourism activity may also come about because of the physical and architectural fabric, especially the dominant presence of urban heritage. Alternatively, the distinctive character derives from an area's linkage to a particular cultural or ethnic group within the city. It is not uncommon for such characteristics to exist in combination like, for example, an entertainment area located within a cluster of historic buildings.

In tourism academic literature, different terminologies are used to refer to such areas. According to Judd,[25] 'tourism bubble' refers to an area in which tourism and leisure are overwhelmingly the predominant activities and in which most buildings and urban spaces are dedicated to leisure use. They typically include a range of facilities such as museums, art galleries, entertainment facilities, leisure shopping, bars, cafes and restaurants. Not only do they offer visitors a typically tourist landscape but they are also isolated from the city's environment and social problems, such as dereliction and poverty.[26] The term was coined at a time – the 1990s – when most tourism activity in cities was concentrated in a few urban areas in which the presence of non-tourism uses was very limited.

Hayllar and Griffin[27] adopt the term 'tourism precinct'. They debate the experience of tourists within a tourism precinct which they describe as: 'a distinctive geographic area within a larger urban area, characterised by a concentration of tourist related land uses, activities and visitation, with fairly definable boundaries'. Precincts may include iconic sights, shopping areas, landmark cultural institutions or places of historical significance. They note that where a number of attractions cluster alongside a range of tourism-related services, these areas take on a particular spatial, cultural, social and economic identity.

Many urban areas are multifunctional offering a wide range of facilities for use by locals as well as tourists.[28] Tourism activity often merges with the daily life of the place to the extent that it is often difficult to distinguish between the two. Tourists and residents make use of the same urban facilities be they shops, catering establishments, cultural attractions and transport facilities.[29] An alternative term for 'tourist bubble' or 'tourist precinct' is 'tourism-oriented areas'. This term is a better reflection of what actually happens on the ground as it acknowledges a high level of tourism activity while at the same time reaffirms that the area is a dynamic part of the city and not something that is distinct from it.

Tourism is not a separate activity that happens only in specific areas of the city. It is much more pervasive. Tourism and leisure activities co-exist with activities and uses that are normally associated with urban areas such as residential and commercial uses. Tourism activity takes place in different parts of the city, engaging with its architectural and cultural fabric. There would be some visitors who are drawn to different areas of the city and who are interested in exploring the city, seeking experiences away from the well-established tourism zones.[30] In large, well-established tourist cities like Amsterdam, London or Paris, city authorities seek to draw tourists away from the traditional core zones towards other parts of the city.

The nature of tourism activity in cities is highly dependent on the size of city. For instance, small- and medium-sized cities with historic central areas tend to have virtually all their tourism and leisure activity concentrated at the city's core. On the other hand, larger cities and global metropolises are subject to rapid urban growth and therefore experience an increasing diversification of tourism space. Newly developed areas are complementary and act as tourist counter poles to the more historic parts of the city.[31] Examples are the newly developed (or redeveloped) districts of La Défense in Paris and Potsdamer Platz in Berlin.

There are different levels of tourism activity in different parts of the city. At one end of the scale, there is the residential area with minimal tourism activity consisting maybe of a few units of shared accommodation or an isolated bar or restaurant that attracts locals and a few tourists. At the other end of the scale, there is a high concentration of tourists within an urban space. Normally this comes about because of an outstanding visitor attraction that most tourists to that city would wish to visit. The area would include numerous shops, bars and restaurants to service the tourist client. Visitor crowding is a common occurrence and such areas are often cited as examples of overtourism. Examples include Fontana di Trevi in Rome, Ponte di Rialto in Venice and Las Ramblas in Barcelona. In most cases, areas with tourism activity in urban areas lie somewhere along this scale between the two extremes.

Do tourists visit a city just for the visitor attractions? Do tourists spend most of their time in leisure settings, relaxing away from their daily concerns? Or do tourists go on holidays only to engage in a particular activity? In many instances, the time spent by tourists in cities involves all three. Apart from visiting attractions, most tourists spend time in leisure settings, free from work-related thought. Some visitors to cities could also engage in particular activities like going to the theatre, playing at a casino or enjoying evening entertainment.

The importance of visitor attractions to tourism destinations is self-evident. This is discussed in detail in the next chapter. Most literature on tourism destination management give particular attention to the role of attractions in city destinations, and rightly so. In doing so, however, one aspect of destinations tends to be overlooked and taken for granted. These are leisure settings. The traditional form of tourism prevalent in cities towards the end of the millennium involved groups of tourists being 'herded' around from one visitor attraction to another, with some break time for shopping or a coffee. Nowadays, the predominant form of tourism is different in that most visitors to cities organise their own travel and accommodation, and more crucially their own itinerary. They decide which places to visit, how to get there and how to spend their time, without being constrained by what is on offer by the tour operator. In this latter form of tourism, the time spent by tourists in leisure settings is greatly increased.

### 4.2.1 Leisure settings

A leisure setting is a place within the city where visitors can slow down, or even rest, before or after visiting an engaging visitor attraction. Leisure settings are also used by city residents, including office workers and nearby residents. A leisure setting would normally be an urban park, a garden, a landscaped urban space or an area overlooking a river or the sea. It could also be a pedestrian urban space that has pedestrian-friendly street furniture such as seating and greenery or a historic area that provides a pleasant ambience for walking and enjoying the historic context. For a place to be a leisure setting for tourists, it has to be relatively quiet and in particular free from air pollution and the tiring noise of traffic. Moreover, it has to be perceived to be safe.

The distinction between a visitor attraction and a leisure setting is sometimes blurred. For example, a park is often used for leisure and relaxation if it is located close to areas of tourism activity. A park could also be an attraction in its own right with tourists visiting it because it is on the city's tourist trail. Visitor attractions, even paid ones, sometimes offer spaces where people can spend time relaxing without actually looking at the site features or artefacts. This is useful because it will extend the time spent by the visitor at the attraction and could potentially result in increased spending at the site.

Water is a common feature in leisure settings. This can be present in many different forms. Often the birth and eventual growth of a city is determined by the presence of a body of water. Many cities are located along a river or lake because of the ready availability of water to sustain human activity. Other cities are located at coastal locations near the sea, often where there is a natural harbour. The history and narrative of many cities is often dominated by the presence of water and the way the water has influenced and shaped the life of the city over the centuries. Water holds a strong aesthetic appeal for many people, especially if there are other features such as historic building or fortifications that add to the visual attractiveness. Very often locations close to the water become sites popular with visitors because of the views. An example of this is a green open space alongside side the River Oder in Wroclaw, Poland. From here, people can enjoy pleasant views of the historic cathedral quarter, Ostrów Tumski, across the river (figure 4.2). The riverside has been re-modelled with stepped seating facing the river enabling people to relax and enjoy the views.

Motionless water can be attractive as it conveys a feeling of calm to whoever cares to observe. Although water itself is a colourless liquid the colour that we see in water is either from the surroundings being reflected on the surface or underwater objects seen through it. In deeper waters, such as rivers and seaports, the water takes on a variation of tones in blue, sometimes also in green. The intensity – or lack thereof – of natural light alters the way that bodies of water are seen and visually experienced. Sunlight adds sparkle and heightens the excitement where there is moving water.[32] In man-made

**Figure 4.2** Views of the historic area of Ostrów Tumski, Wroclaw, from across the river.
*Source:* Author's own.

water features where water is in motion, the use of bright artificially lights can create remarkable effects. Moving water such as weir overflows and waterfalls are interesting to look at and often generate interest from visitors as they greatly enhance leisure settings, visually and acoustically. The attraction of moving water is often capitalised in the design of fountains. This is the case at Fontana di Trevi in Rome where a small square is occupied almost entirely by an elaborate fountain. This has become a major tourist attraction not least because of the play of moving water over the Baroque sculptures of the fountain.

Curtis[33] describes the visitor experience that the Thames offers to visitor and tourists. His description is applicable, to varying degrees, also to rivers that pass through or close to the commercial or historic centre of a major city, especially those that are oriented towards tourism.

> The river can be a place for reflection and contemplation, offering riverside walks and traffic-free spaces to take in views; and it provides numerous attractive places for eating, drinking and socialising; ultimately it provides diverse views and widescreen vista sand pleasing setting for the numerous cultural and historic sights along its course.

Curtis also refers to the Thames as a waterway that offers a means of transport between the sights. Water transport provides visitors the opportunity to view the city sights from a different perspective and in a leisurely manner.

## Case Study AR1: Three examples of the creative use of water in leisure areas

### Multimedia Fountain in Szczytnicki Park, Wroclaw, Poland

Wrocław Multimedia Fountain (Wrocławska Fontanna Multimedialna) is set in the pleasant surroundings of Szczytnicki Park at the outskirts of the city. It was inaugurated on 4th June 2009 on the occasion of the twentieth anniversary of the first free elections in the postwar Poland.[34] It is claimed to be the biggest fountain in Poland and one of the biggest in Europe. The water basin contains three hundred water jets which spout water in many different ways such as jets, mists and spurts. A spectacular show is created by means of water effects with frontal and backlighting from eight hundred LED lights and synchronized to music. Three-dimensional images are projected onto the water screen. Special shows are designed using images and even short film projections to commemorate special events and important dates of Poland's history. After dark, the shows are especially impressive as the colours of the water, light and projections appear more vibrant. Minutes before a show is due to start crowds gather waiting in eager expectation. The fountain has become a must-see attraction for anyone visiting the city of Wroclaw. Szczytnicki Park is visited by many domestic and international tourists, not only because of the Multimedia Fountain[35] but also because of other attractions namely the historic Japanese Garden and the Centennial Hall.

### Miroir d'Eau in Bordeaux, France

The Miroir d'Eau (Water Mirror) in Bordeaux[36] is a reflecting pool made of granite slabs covered by two centimetres of water. At an area of three thousand square metres it is the world's largest reflecting pool.[37] A water spray system allows it to periodically create mist in warm season. It is located along the Garonne River just across from Place de la Bourse, a palace built in the mid-eighteenth century. It alternates between a mirror effect and artificial misting to create an extraordinary effect, greatly enjoyed by visitors. In summer, it is a place where children enjoy walking and playing among the water jets.

### Princess Diana Memorial Fountain, London, UK

The Princess Diana Memorial Fountain occupies an area of green landscaped ground sloping down towards the calm water of the Serpentine pool in Hyde Park. The design of the fountain is in marked contrast with many public monuments and memorials.[38] Rather than being an obtrusive feature imposed onto the landscape, the fountain's sculptural form integrates into the natural slope of the land. The fountain is shaped as a large oval stream bed made of granite with water pumped to the top of the oval and allowed to flow down either side.

On one side, the water flows fairly smoothly to the downhill end of the oval. The other side consists of a variety of steps, curves, and other shapes so that the water plays in interesting ways as it flows down to the tranquil pool at the bottom. The two water flows are intended to represent two sides of Lady Diana's life: happy times and turmoil.[39] At a cost of £3.6 million, it was inaugurated in 2004. The memorial fountain met with some initial difficulties largely because parts of it were found to be unsafe for playing children. The granite stream bed is fairly shallow and has a varying width of three to six metres. There is a distance of about eighty metres between the highest point of the oval, where the water is pumped out, and the tranquil pool at the lower end.[40] People engage with the memorial in different ways. Many people sit on the granite edges of the tranquil pool at the lower end. Some lay on the grass within the oval or just outside it. Many watch the water as it flows down the granite stream bed. Whether flowing or still, the water provides an excellent medium for children's play. People enjoy pleasant views of the Serpentine and of Hyde Park's lavish green landscaping as they walk around the memorial fountain, or are simply relaxing on the granite edges or the grass. The overall design makes clever use of water to create an ideal leisure setting.

### 4.2.2 Tourism in city districts

Larger cities feature various areas or districts with different characteristics and different levels of tourism activity. Tourism activity takes place in city districts which can be termed as cultural, historic, ethnic, entertainment, support or regenerated districts.

Cultural districts represent areas of both cultural production and consumption, characterised by a variety of cultural venues and events. Very often cultural districts evolve over time involving an organic and subtle facilitation of cultural activities, combined with an appropriate built form.[41]

In many cities, the historic area is a main attraction and an important part of the city's tourism product. Academic literature dedicated much attention to the historic area's role in tourism following the seminal work on the tourist-historic city by Gregory Ashworth and John Tunbridge published in 1990.[42] Clusters of well-preserved historic buildings set the area apart from the contemporary and mundane built environment of the surrounding city. This is the basis for the area's attractiveness, often augmented by a broad range of touristic uses that are accommodated within the historic fabric.[43] Part of its appeal is the socio-cultural character of the place and tourists would often wish to get a glimpse of the residents' way of life. Chapter 9 discusses tourist-historic cities in the broader context of urban heritage and conservation.

Large cities across the world, including Europe, are becoming increasingly diverse because of historic and contemporary international migration patterns.

Ethnic quarters have often been viewed with a degree of mistrust. In many cities, they are living evidence of past migrations of ethnic minorities.[44] For decades these urban areas and their inhabitants were relatively isolated from the larger community and served as a refuge from the dominant society. The diversity of urban population is a factor increasingly emphasised as a creative resource for cultural, social and economic development.[45] The trend of inward migration continues in many European cities because of the growing mobility of people. It is not uncommon for people of a particular ethnic origin to take up residence in a particular part of the city. Over many years, this gives rise to an ethnic district with identifiable features reflecting the cultural background of the ethnic group. The cultural landscape of many cities evolved over time with different ethnic groups integrating into the city's social fabric. In some cities, these have evolved into tourist attractions. Inclusion of the ethnic precincts in the modern city add to the richness of the city's experiences by providing opportunities to 'dip' into other cultures through unique food and beverage experiences, shopping and participation in ethnic festivals.[46] Ethnic populations and the expression of their culture, rituals, customs and cuisine help build a cosmopolitan feel to the city, one that is consumed by tourists. The placement of culturally specific features, like Chinese language street signs, provide the tourist and city residents with an experience that is different and hints of the exotic without having to travel far afield. One example is the Balti Triangle in Birmingham. The city lured immigrants from all over the world to work in the car factories at a time when the UK's automobile industry was at its peak. Birmingham continued to maintain a large ethnic population thereafter making it one of the most culturally diverse urban areas in the UK. Birmingham's promotional material encourages visitors to search for the Balti Triangle in south Birmingham where they can enjoy many restaurants serving traditional food and shop for ethnic clothing and jewellery.[47] Some cities choose to invest in the upgrading of the physical landscapes and general built environment. This is done, in part, in an effort to use heritage and ethnicity to attract visitors. Leisure and tourism are closely connected to these revitalised areas; giving locals the opportunity to financially benefit from the increased attention. According to Spirou,[48] 'Cultural diversity can be celebrated, repackaged and presented as an asset in the quest for attracting visitors' and refers to Greektown in Toronto and Chinatown in Montreal, Canada as examples. These received extensive makeovers as part of their tourism-based regeneration strategy.

Former warehousing areas are often considered favourably for conversion into entertainment districts. Having been abandoned, these areas would be in need of a major revamp and a new financially viable use. The city authorities would need to embark upon significant planning rethink for the area but even so, compared with redevelopment, former warehousing areas can be converted relatively quickly for entertainment-related uses. An added advantage of these areas is that they can accommodate large crowds.[49] An interesting example of this is Liverpool's waterfront development discussed in Case Study AR2.

In cities that are popular with tourists, a common occurrence is the build-up of what can be termed as a support district. An urban area of limited intrinsic tourist interest changes and evolves over time because it is located within walking distance of a popular tourist attraction. A support district is the most convenient location for hotels, catering establishments and shops to service the many visitors to the attraction either before or after their visit. One example of this is Plaka in Athens, Greece. It is a quaint and picturesque neighbourhood, located at the foot of the hill on which the Acropolis is located. Plaka serves as the gateway to the millions who annually visit the impressive archaeological remains at the Acropolis. In recent decades, it has evolved considerably gaining a strong tourist element and yet retaining its strong residential heritage within archaeological grounds.[50] It maintains a vibrant commercial, cultural, residential and tourist identity. While the urban character projects a bygone era of Greek society and culture, it has been subject to widespread commercialisation.[51] This raises questions of authenticity and is detrimental to the qualities that make it popular with tourists.

Since the latter part of the twentieth century, there have been many instances of cities regenerating areas that have become run down or derelict. New buildings are constructed and some historic buildings are renovated. In such instances, the district appears to be manufactured rather than having evolved organically based on some specific characteristic of the place. The redevelopment of redundant commercial waterfront areas for tourism and recreation is a common feature in many cities.[52] The waterfront location is often an attribute that makes such areas highly suited to tourism uses.[53]

## Case Study AR2: Liverpool's waterfront development

In the 18th and 19th century, Liverpool docks and waterfront warehouses were bustling with activity as the city was an important interchange between land and sea transport for the trading of goods and commodities.[54] Early in the 20th century Liverpool port activities were in decline as the transport and storage of goods were transferred to more modern port facilities nearby. In the latter part of the 20th century Liverpool felt the full effects of deindustrialisation and broader manufacturing decline. Many of the docks were closed to shipping and warehouses were left derelict, with some eventually being demolished. The Royal Albert Dock was closed to shipping in 1972.

In the early 1980s, the Merseyside Development Corporation was set up and tasked with redevelopment of the area, in partnership with the private sector. The overall objective was to bring about improvements to local conditions. The regeneration project, spread over many years, proved to be very successful as it transformed the area into a place to visit and a sign of the city's economic rebirth and waterfront revival. The Royal Albert Dock became a popular heritage attraction as well as a leisure and entertainment hub.

The success of The Royal Albert Dock spurred on investment in Queen's Dock and adjoining docks and wharfs. A cruise passenger terminal was built at a revitalised Pier Head enabling more people to visit the city, this time from the sea. Regeneration also involved residential developments on quays alongside the Liverpool Marina. In 1998, the Crowne Plaza Hotel was part of a major redevelopment of Princess Dock. Later developments in the area involved mixed uses combining residential with office developments and hotels. The regeneration project converted the entire area from a derelict and unwanted part of the city to a dynamic urban area with a range of leisure, commercial and residential uses as well as becoming a tourism district in the city.

*Chapter 5*

# City Resources for Tourism (2)
## *Visitor Attractions*

Visitor attractions have a crucial role in the development and success of tourism destinations. As discussed in a previous section, there are several resources that must be present for a city to be effective as a tourism destination, but it is visitor attractions that play the most important role. The purpose of this section is to better understand the nature of attractions, to introduce the variety of visitor attractions and to consider their varying role in differing city destinations.

Most people tend to think of city destinations in terms of their most wellknown attractions. They travel to destinations to see and experience these attractions. It is generally the attraction that prompts the tourist to travel to the city destination in the first place. This could be a monument, a museum, a historic building or area, a church or cathedral, an architectural icon or a shopping street among other things.

Defining a visitor attraction is no easy matter. The concept of 'attraction' is a very broad one, encompassing many different sights and sites. Any site that is of sufficient interest for people to visit it could be deemed to be a 'visitor attraction'.[1] According to Keyser,[2] an attraction is 'a designated resource which is controlled and managed for the enjoyment, amusement, entertainment and education of the visiting public'. This definition is somewhat restrictive as it applies to facilities that have an established management structure and are open at established times for visitors, normally against payment. It does not cover urban spaces and features (e.g. historic areas, pedestrian streets, parks and memorials) that are maintained and managed by the city authorities as a public good for the use of residents as well as visitors. A more comprehensive definition of a visitor attraction would be a site that is of sufficient aesthetic, narrative and cultural interest to provide for the enjoyment, amusement, entertainment and education of visitors.

In discussing visitor attractions, it is important to include also those features, sites and areas for which no payment is needed to view or visit. For example, the Upper Barrakka Gardens in Valletta, Malta, provide fantastic views overlooking the Grand Harbour. Visitors can freely walk into the gardens to enjoy the views. The site is one of the most popular places to visit in Valletta.[3] Therefore, it should be considered as part of the tourism product and also as a visitor attraction, despite not requiring payment for access. The same can be said of urban areas which are visited by tourists because of their historic, cultural or ethnic interest.

Many attractions are not purpose-built in that they were not designed and built specifically to appeal to the tourist. For example, many historic town centres and fortified towns grew and were developed over time for a purpose other than tourism, even if today they often become important visitor attractions. Cathedrals, churches, historic buildings and public gardens were not built to bring in tourists but over time some of them have become interesting to visitors for architectural or historical reasons or because of narratives that are linked to them. Although not purpose-built for tourism, many sites were adapted to handle visitors. One example of this is when a visitor centre is built close to an important archaeological site. Another example is an important historic building (e.g. a stately home, castle or cathedral) that is provided with reception, ticket sales, signage, interpretation and other elements that are necessary for visitor management. Apart from pre-existing buildings and sites, the more important city destinations also include attractions that were developed specifically to cater for visitors and for the tourist market. Examples include museums, art galleries, heritage centres and theme parks.

In many attractions, the increasing need to generate alternative revenue streams has led to the expansion of the core activities of the attraction. New attractions integrate some element of retail and catering, whereas existing ones adapt to create revenue-generating spaces.[4] Spaces for conferences or for the hosting of receptions are also useful.

A useful classification of visitor attractions is as follows.[5] A primary visitor attraction is one which influences a tourist's decision to visit the city. This requires information received by a tourist in some form, normally over the internet. The information interacts with the person's needs and aspirations to motivate a visit. A secondary attraction is a feature or characteristic that is known before a person visits but is not sufficiently significant to shape motivations. A tertiary attraction is unknown to the tourist prior to the visit but maybe discovered as and when the tourist explores the destination. Analysing attractions at a destination by primary/secondary/tertiary classification can help describe and explain patterns of behaviour of tourists. This, in turn, can be made use of for purposes of marketing, management and planning of the city destination.

There are many visitor attractions that are in open countryside but which can easily be reached from a nearby city. Many city destinations offer half-day or full-day coach tours to attractions that are within a one- or two-hour drive. In the minds of tourists, these out-of-city sites are as much part of the city's tourist appeal as other attractions that are within the city itself. Examples of these sites include rural villages, stately homes, castles and monasteries. There are also out-of-city attractions that are based on natural resources such as national parks, nature reserves, wildlife parks, countryside trails, vineyards, natural coastlines and sandy beaches.[6] Nature-based resources are also relevant to city tourism either because they are located very close to the city or because tours are organised from the city to remote natural attractions. Tours out of the city are a useful means for enriching a city's product offer to tourists and hence provide for added value to the overall experience of the city destination. For a city with limited attractions, this is essential if it is to compete in the tourism market.

Sometimes the cultural significance of a site or attraction is of such international importance that it is referred to as a 'cultural icon'. This may be a historic building such as a castle or monument or it may be a building or structure of particular visual or historic interest such as a tower, bridge or pier. Contemporary architecture could also be a cultural icon in which case it would be referred to as an architectural icon – something that is discussed further in chapter 10. In several cities, a cultural icon, or a synthesis of cultural icons, act as a magnet to attract the tourist. The most successful city destinations are those fortunate enough to have an internationally known dominant icon. The fame of the cultural icon often extends far beyond the city or country itself with the result that the image of the icon and the destination become inseparable in the mind of the prospective visitor. Examples of cultural icons include the Tower Bridge in London and the Eiffel Tower in Paris. None of these buildings were created with the deliberate intention of attracting tourists. Over time, however, 'their appeal has widened to a point that tourism flourishes because they exist'.[7]

A cultural experience takes place when a visitor seeks to know, learn or understand something about the culture of the destination. The cultural tourist seeks to engage emotionally and cognitively with places, people and aspects of lifestyle.[8] Sightseeing does not constitute a cultural experience, unless the visitor seeks to learn and understand knowledge and information related to that sight. The various types of attractions discussed below offer a cultural experience in some shape or form, because they enable the visitor to learn and understand better some aspect relating to the place or to modern society.

## 5.1 MUSEUMS AND SITES OF CULTURAL SIGNIFICANCE

### 5.1.1 Museums

Various social forces have influenced museum development in the past 200 years. At the beginning of the nineteenth century, the public began to gain access to collections held in private ownership that were previously opened only for private viewing. This became increasingly common and over the next century, museums began to be established for public viewing for the benefit and education of the public. Museums changed from private collections symbolising personal wealth to educational institutions for the wider public.

For many city destinations, museums are a vital element of the product that is offered to tourists. The traditional role of a museum is the collection and conservation of artefacts of artistic, cultural, historical, or scientific importance. In modern societies, economic planners and tourism stakeholders perceive a museum's role as an attraction and a means to attract visitors to the city. Various other purposes are attributed to museums including serving as recreational facilities, centres of culture in the community, scholarly venues or educational resources. Some museums are seen as a means for the promotion of civic pride or nationalistic endeavour. Given such a variety of purposes, museums reveal remarkable diversity in form, content and even function. Yet, despite such diversity, they are bound by a common goal: 'the preservation and interpretation of some material aspect of society's cultural consciousness'.[9] In museum curatorship it is not uncommon for tensions to develop between these different roles.

Since the 1990s, there have been significant investments across Europe in the cultural infrastructure of cities with new purpose-built museums and galleries, major building refurbishments and extensions. Many of these investments received extensive media coverage. These developments have drastically improved the standards of museum facilities, the quality and variety of displays and the media used for interpretation. This was also a period of fundamental change in the design and shaping of museums which, according to Macleod et al.,[10] brought about

> what might be called 'narrative environments'; experiences which integrate objects and spaces – and stories of people and places – as part of a process of storytelling that speaks of the experience of the everyday and our sense of self, as well as the special and the unique.

These changes were driven by funds made available by national governments and city authorities, as well as advances in digital technologies and a shared understanding of the museum as a storyteller. This has enabled museums to offer engaging, meaningful and memorable experiences to a broader range of visitors.

Chapter 10 describes how museums have moved away from being elitist to becoming popular places of leisure for a wider audience. More creative approaches to marketing is one reason for this trend. Another is the architecture.[11] The increased popularity of museums has implications for tourism activity at city destinations as tourists are more likely to spend time visiting museums. The more well-known and popular museums are the ones most likely to bring in the crowds.

Museums are not site specific. Almost any town or city can invest in its tourism product and establish a museum by assembling and presenting collections of locally significant artefacts. Museums can range in scale from high-profile, internationally known institutions to lesser-known city sites and small community museums in secondary cities. The largest museums are located in major cities, while thousands of local museums exist in smaller cities, towns, and rural areas. The most popular and possibly most common museums are those dedicated to art. There are also museums of natural history, archeology, maritime, science, history, war and other topics. Museums differ widely in the ways items are selected, displayed and interpreted. This has implications on market segmentation and marketing. Recent museum trends favour hands-on interactive interpretation as a way of actively involving the visitor.

Museums can be housed in many different types of buildings. These can be purpose-built in which case the overall layout and design will be tailor-made for requirements. Alternatively a museum can be housed in an existing building. Normally this would be a building of historic significance that would have to be adapted to cater for the museum requirements.[12]

Across countries, the rules regarding free versus charged admission differ. In most European cities, admission to a museum is against the payment of an entrance fee. This is understandable given that their main source of income is from entrance tickets. Moreover, city authorities and national governments are normally reluctant to commit themselves to provide financial support to museums over the long term. In some countries, museums have an admission fee but then concessionary rates are offered to different sub-groups of people such as the elderly and under eighteen. Some museums offer free admission to those subscribing to membership. Rules regarding admission fees change over time according to the prevailing political view. 'Market-orientated' governments tend to consider museums as commercial entities with a duty to compete in the market. 'Welfare-orientated' governments are more prone to favour free entrance to museums consistent with their social and cultural obligations towards the community.[13]

There are also museums that offer free admission for all often with a plea for voluntary contributions. In 2001, the British government decided to eliminate entrance fees to England's national museums. Government-sponsored museums saw visitor rates more than double in ten years with almost 18 million people visiting the thirteen attractions in 2010/2011 compared with 7 million in 2000/2001.[14] The justification for free museums is a social and cultural one as summed up by a prominent British government minister in 2011[15]: "It was about opening up the great institutions, making history, and art and culture, available to the widest possible number of people – and bringing a sense to everyone that these were possessions that we all owned and shared." One argument against free admission is that it is harmful to those museums and visitor attractions that do charge. A representative of Britain's independent museums pointed out that museums are very expensive organisations to run and manage and that their venues lose out as people choose to go to free museums. He argued that more than 50 per cent of British museums are independent and rely on revenue from admissions to survive. Free admission to national museums creates a market distortion and a misunderstanding among the visiting public.[16]

In 2014, the Italian government opted for free admission on the first Sunday of every month to all state museums and monuments. This has resulted in consistent increases not only for free visits but also for those which were charged. In 2016, free visits to Italian museums increased by 9 per cent compared to the previous year, whereas revenue increasing by 12 per cent. Evidence suggests that the policy of promoting free admission to museums for one day per month has encouraged more fee-paying visits during the rest of the month.[17] From a policymaking perspective, Cellini and Cuccia[18] claim that the free-admission policy to public museums is beneficial not only on subsequent charged visits to museums but also to the culture industry and the tourism industry.

The pricing policies of state-owned museums is also relevant to tourism. These are a vital part of a city's tourism product. Being state-owned, they tend to have had more public funding over many years and therefore they normally offer the most interesting and culturally rich attractions in a city. Free admission makes these museums more accessible to a much wider audience of tourists, presumably resulting in an enhanced tourist experience of the destination. On the other hand, funding is an issue that needs to be addressed. There cannot be free admission to tourists without significant state funding. This is a policy option that national government and city authorities should actively consider.

Musée du Louvre is the most visited museum in Europe and the world. Foreign tourists comprise 75 per cent of visitors. Like many high-profile museums, Musée du Louvre often had problems of severe overcrowding at the more popular artworks and at peak periods. To address this issue and improve the quality of the visitor experience, the museum introduced a new visitor flow management programme and online booking system. This resulted in a 20 per cent reduction of visitor numbers during the peak summer months and a marginal reduction in the overall yearly figures. In 2019, 9.6 million people visited Musée du Louvre, compared to 10.2 million in 2018.[19] Figure 5.1 lists the twenty most visited museums in Europe.[20]

### 5.1.2 Palaces, stately homes and historic houses

There is a form of historic attraction in which the building and its landscaped grounds are an important reason for the visit, even if the artefacts and features displayed within the building itself are worthy of being viewed and appreciated. In Renaissance Europe, from the sixteenth century onwards, royal and privileged families built their lavishly adorned palaces in prominent locations in the city (figure 5.2) as well as their retreats and hunting lodges in the countryside. Many of these palaces are now attractions in which visitors enjoy being amazed by the opulence of the richly adorned interiors, as well as the furniture and the furnishings. They also enjoy getting to know about the way of life of the former royal residents and interesting narratives and anecdotes about their lives.

### 5.1.3 Castles, fortifications and fortified towns

Throughout history communities, groups and individuals felt the need to build structures to defend their settlements and themselves from the attack of invaders. Defunct military buildings and structures are a plentiful and widespread resource across Europe. In its most basic form, the military architecture consisted of a single building or small group of buildings occupying a relatively small area and surrounded by high defensive walls. This is referred to as a 'castle' or 'fort'. In some contexts, 'castle' could also refer to a sizeable area defended by walls and occupied by a significant number of

| Rank/ Museum/ City | Attendance 2019 | Rank/ Museum/ City | Attendance 2019 |
|---|---|---|---|
| 1 Louvre, Paris | 9,600,000 | 11 Centre Pompidou, Paris | 3,270,000 |
| 2 Vatican Museums, Vatican City | 6,883,000 | 12 Science Museum, London | 3,254,000 |
| 3 British Museum, London | 6,208,000 | 13 Museo Nacional del Prado, Madrid | 3,203,000 |
| 4 Tate Modern, London | 6,098,000 | 14 State Tretyakov Gallery, Moscow | 2,836,000 |
| 5 National Gallery, London | 6,011,000 | 15 Rijksmuseum, Amsterdam | 2,700,000 |
| 6 Natural History Museum, London | 5,424,000 | 16 La Cité des Sciences et de l'Industrie, Paris | 2,370,000 |
| 7 State Hermitage, St. Petersburg | 4,957,000 | 17 Galleria degli Uffizi, Firenze | 2,362,000 |
| 8 Reina Sofia, Madrid | 4,426,000 | 18 Auschwitz-Birkenau Museum, Oświęcim | 2,300,000 |
| 9 Victoria & Albert Museum, London | 3,921,000 | 19 National Museum of Scotland, Edinburgh | 2,210,000 |
| 10 Musée d'Orsay, Paris | 3,652,000 | 20 Van Gogh Museum, Amsterdam | 2,100,000 |

Figure 5.1   The most visited museums in Europe in 2019.

**Figure 5.2 The beautifully adorned loggia of the Castello del Buonconsiglio, Trento, Italy.** This fortified palace was home to Trento's bishop-princes until 1801. *Source:* Author's own.

buildings. Hence, the dominant defensive structures overlooking Edinburgh are referred to as 'Edinburgh Castle'. The word 'fortifications' refers to defensive walls which normally form part of a larger network of defensive structures to defend a sizeable area of territory, including buildings. In combat situations, a line of fortifications on its own did not defend a town but a line of fortifications in combination with other fortifications and defensive structures did. 'Fortified landscape' is a term that is sometimes used in some academic literature.[21] This is a better reflection of what actually happens on the ground because, besides buildings and other structures, many defensive systems include engineering works, earth structures, ground clearance and excavated ground. A town defended by a wall or a network of fortifications is referred to as a 'fortified town'. An alternative term that is sometimes used is 'walled town'. In tourism and urban conservation literature, readers are sometimes confused because of the different terminologies that are used for defensive military structures. This paragraph provides some clarification of the different terms that are used.

Extensive walls and fortifications around towns came about because of a long history of tensions and conflict between neighbouring peoples. In many towns and cities, these walls were partially or completely lost either through neglect or, more frequently, because they were removed to make way for much-needed development of houses and roads. There were however many others which remained intact and virtually untouched by inappropriate interventions.

Fortified towns have features that make them very suitable for commodification into tourism products. Fortifications are distinctive features that add to the town's uniqueness while at the same time making it more easily legible.[22] A wall around a historic town is significant in a number of ways. It tells a story; this is a town that was attacked or that was liable to be attacked.[23] It is a reflection of the science of warfare which prevailed at the time it was built. The fortifications surrounding the town make for a compact area with a clearly visible edge or boundary, thus clearly physically defining the town. This allows for an easy and quick understanding of the town plan by even the most transient of visitors. For most fortified towns, the street layout can be easily visualised and remembered, and an instantly usable mental map can be formed.[24] Thus, circulation is easier and the chances of being lost in an unfamiliar place are reduced.[25] This reduces visitor anxiety as it reduces concern about getting lost. Where still fully intact, fortification walls provide a clear definition of the extent of the historic area. Walls are a unifying factor of the historic town's urban self-image.

The presence of physical features, such as hills, river banks or harbour sides, within or associated with the walled town renders it unique and, in some ways, exceptional. Examples of fortified hilltop towns include San Gimigniano (Italy), Granada (Spain) and Elvas (Portugal). From the surrounding countryside, the views of these towns are remarkable because of their elevated position and the bulkiness of the massive surrounding walls. Rhodes (Greece), Piran (Slovenia), Dubrovnik (Croatia) and Hellevoetsluis (Netherlands) are examples of fortified towns adjoining rivers or harbours. Many of these examples are world heritage sites. In the Maltese Islands, there is a unique situation in that there are four different fortified towns within the islands' small geographical area. Mdina and Cittadella are located on high ground at the centre of the islands of Malta and Gozo, respectively. Another two are located facing onto Malta's main harbour, the Grand Harbour, namely Valletta and Cottonera. The latter is a complex of fortifications over an extensive land area within which are three small towns Vittoriosa, Cospicua and Senglea (also known as Birgu, Bormla and Isla, respectively). These various fortified places have a long history of residential occupation and continue to operate as urbanised settlements to this day.

There is a curious paradox about fortifications. They are admired by many for the aesthetics, the construction, their scale and for their symbolic meanings. If well maintained and presented, they provide for the enjoyment of visitors. Historically the construction of interesting geometrical shapes in rock and stone had one intention – that of facilitating the killing of attackers. For this reason, Ashworth and Bruce[26] argue that military heritage is inherently 'dissonant' in that it can potentially evoke feelings of disquiet and even repulsion. Many visitors to fortifications have little consciousness of the original brutal function and thus suffer little or no psychological disturbance. The lapse of time may blunt the brutal violence and render the purpose of fortifications quaint and even romantic.

In many cities, the historic core plays a crucial role in tourism in a number of different ways. Normally the historic core is the location of the more important attractions. These could be churches, museums, historic buildings or some other form of cultural attraction. Many tourists spend time walking and exploring the historic area thus making it an attraction in its own right.[27]

### 5.1.4 Archaeological sites

Archaeological sites are popular tourist attractions at many destinations. Archeological sites are normally associated with relics of the past that are excavated from the ground, although it is not uncommon for archaeological sites to include above-ground structures. Archaeological sites are fragile resources so the more popular sites may be negatively impacted by large numbers of people. Hence the need for careful visitor management which may include the restriction of public access to parts of the site.

The provision of site interpretation and appropriate facilities enables visitors to appreciate the cultural and archaeological value of the site. Sustainable enjoyment of an archeological site secures income from entrance fees and thus contributes to the site's long-term preservation. From the visitor's point of view, the downside of a visitor centre is that it decontextualizes the experience away from the actual archaeological site. In some sites, providing visitor facilities and interpretation could be difficult largely because of the need to respect the sensitivity of the site and its landscape setting. It might be possible however to provide a visitor centre at some distance from the archaeological site itself and thus keep the impact on the context to a minimum. This was the case with the visitor centre built at the Ħaġar Qim and Mnajdra temples in Malta.

It is not uncommon for archeological remains to be found when excavating a site for the construction of a new building or road. Depending on the importance of the find, the authorities may decide to carry out a detailed archeological study of the site and then allow for the removal of the remains. For more important discoveries, the plans of the building or road may have to be changed to allow for the retention of the archaeological remains in situ. One option is to construct the building on piles or columns so that the underlying archaeological remains are left largely undisturbed. One example is Bloomberg's European headquarters in London which was built above the Roman-era Temple of Mithras. Visitors descend through steep stairs to seven metres below the city streets to view the remains of the temple and other ruins dating back to about 250 AD.[28] The site is now a visitor attraction, the London Mithraeum, in the heart of London.

### 5.1.5 Industrial and transport attractions

The industrial revolution in the eighteenth century brought radical changes to society across Europe. New products and materials were produced, while radical improvements in public transport infrastructure enabled people to move more freely between cities. Industrial heritage refers to the buildings, machinery and other artefacts that throw light on the forces of change during those times. The industrial revolution originated in Britain. Over the past half century, there has been renewed interest in the many redundant buildings and obsolete machinery dating back to this period.

Many of the early industrial buildings and structures were also architectural gems in their own right and this gave impetus to the drive to preserve these buildings and restore them for tourism. A well-known example is Ironbridge Gorge, near Birmingham, England, where the industrial revolution is said to have started in the early 1700s. Ironbridge is the world's first bridge constructed of iron and it had a considerable influence on developments in the fields of technology and architecture. It is now a UNESCO World Heritage Site. There is a wide range of industrial sites with tourism potential. Several coal mines in Wales have been converted into tourist attractions. Textile mills in North England were pushed out of business by the import of cheap textiles in the 1950s. Some of these mills have taken on new life as museums[29] or even as living history museums as is the case with Black Country Living Museum (Case Study VA1).

### Case Study VA1: The Black Country Living Museum, near Birmingham

The museum occupies a 10-hectare site at the outskirts of Birmingham, England. It recreates the industrial landscape that had first emerged in the 1830s. Houses, shops, workshops and public buildings were dismantled and rebuilt to create an early 20th-century village. The village preserves a cross section of social and industrial history. Visitors are able to see the recreated streets, small shops, a school and a chapel. Activities in the buildings are demonstrated by staff in period costume. They are also able to get a glimpse of Black Country's industrial past in a colliery, a coal mine, a lime kiln, a small brass foundry, various workshops, a canal with canal boats and a working boat yard. The attraction also includes a transport collection with old trams and trolley buses.[30]

Interest in industrial heritage spread to include interest in modern industry. Several companies recognise the potential for good public relations. They open their doors to the public to see work in progress in the factory. Alternately they maintain a workshop or a small exhibition on site. Apart from public relations, watching the production processes encourages visitors to purchase the product and souvenirs of the product. There are also production processes that are more craft-oriented and are on a much smaller scale. One example of this is Murano glass. Tourists in Venice are offered free boat rides to the island of Murano for glass blowing demonstrations inside a traditional factory. The expectation is that visitors will buy products before they leave. Some international companies have developed visitor centres or attractions to display all aspects related to their product including production, marketing and consumption.[31] Examples include Cadbury World in Birmingham, Guinness Storehouse in Dublin and Swarovski Kristallwelten in Innsbruck.

Transport is another form of technology that generates significant interest. Transport museums are popular attractions in several cities. Many transport museums are focused on railways or on trams. Vintage cars are the subject of another type of transport museum that is popular across Europe, not least because of a large number of car and vintage car enthusiasts. The same can be said about aviation museums. There are also museums devoted to early carriages, bicycles and canal boats. Where the displays are too large to house indoors, there are high costs for the conservation of artefacts. This makes the financial viability of the museum problematic.[32] In some cases, old vehicles are restored and brought back into service for leisure purposes. One example is steam railways that run through scenic countryside in Wales and in other parts of Britain. Canal barges are another form of leisure transport, made possible in Britain because of 2,000 of navigable inland waterways.

### Case Study VA2: The Guinness Storehouse, Dublin

The Guinness Storehouse in Dublin[33] is housed in what was previously an old abandoned fermentation plant in the middle of the Guinness brewing complex in the industrial part of Dublin. The building encompasses fifteen thousand square metres of floor space spread over seven floors surrounding a huge central atrium. Visitors have the possibility of learning about the product and about the company's two hundred and forty-three-year history. More than that the Storehouse is used as a stage where visitors can co-create their own experience. At the 'Guinness Academy' visitors are taught how 'to pour the perfect pint' and are awarded a certificate for their efforts. A memory wall is provided where visitors can leave a message or spend time reading the messages of others. One area is dedicated to the brand's print, digital and TV promotional campaign over the years, giving visitors the opportunity to interact with some of the adverts. The Storehouse journey ends in the rooftop bar from where visitors can enjoy a 360-degree view of Dublin. The Storehouse is also an event venue with training facilities for staff and spaces for conferences and art exhibitions.

The Storehouse uses ultramodern facilities to breathe new life into the brand and reconnect the company with a younger audience. The attraction provides a special atmosphere and experience by telling a story of an old brand using modern information technologies. The experience is further reinforced by the symbiosis of an old industrial building and its modern architecture conversion. In an interesting analysis of the Guinness Storehouse and its different sections, Müller and Brunner-Sperdin[34] explore how the staging of experience relates to Pine and Gilmore's[35] four realms of experience namely entertainment, education, aesthetic and escapist.

### 5.1.6 Features of local culture and way of life

Tourists are increasingly seeking places that offer them an insight into the daily lives of the residents of the cities they visit.[36] For example, a study on Valletta[37] found that tourists were fascinated to see people from their upper-floor apartment windows lower a shopping basket to buy bread from the street hawker. Somewhat surprisingly, one study respondent said he took photos of people doing chores like dusting the house front door or sweeping the pavement in front of their homes. The locals' way of life is intangible and is therefore difficult for tourists to observe or to 'capture' during a short visit. They therefore look out for features that reflect local life. In the same study on Valletta referred to previously, tourists observed facade features such as the timber doors, the door knockers and the house names, seeing them as a reflection of the persons living within. Others noted features inside churches – religious features that they considered to be indicative of the way of life of the community. Churches are discussed elsewhere in this chapter. In cities, other buildings that provide insight into the lives of local residents are food markets. These are discussed in the next section.

### 5.1.7 Food markets in historic city centres

Food markets provide some insight into the lives of local residents. They are representative of an urban area's culture as reflected in the tastes and eating habits of its residents. By consuming locally sourced food products, tourists associate themselves with local values and local identities. The experience allows tourists to enjoy the city more as if they were local citizens. In this sense, one way of behaving as a local is to visit a food market.[38] Food markets located in historic cores are notable for their urban heritage and architecture. They are part of the tangible urban heritage and also, because of the food culture, of the city's intangible heritage.

In an interesting study of food markets in Madrid and Barcelona, Crespi-Vallbona and Pérez[39] analyse different food markets and categorise them into three categories. The categories are based on the extent to which the markets have been transformed because of tourism. The first category is referred to as 'touristified markets'.[40] Examples include San Miguel in Madrid and La Princesa in Barcelona. These have been redesigned and converted into food halls to meet tourist demand. They are frequented almost exclusively by tourists. Apart from the sale of foodstuffs, the food halls include bars and restaurants and periodically stage culinary events.

Another type is the food market that attracts the resident population as well as tourists. Despite being traditionally oriented, or maybe because of it, these markets are interesting to tourists and are also promoted in tourist literature. Crespi-Vallbona and Pérez note a certain tension in these spaces as the traditional food retail for the residents co-exists with the modern servicing of tourist needs. An example of this is La Boqueria market in Barcelona. The current building dates back to the nineteenth century, although some remodelling was carried out in 2000. It is along Las Ramblas, a wide street and promenade that is very popular with tourists. It is inevitable therefore that tourists walk into La Boqueria, not least to explore an aspect of the life of the city. The food market includes many traditional stalls displaying and selling meat, fish, fruit and vegetables, cheeses, bread and other quality food products. There are also plenty of stalls that target tourists. They sell fruit salads, chocolate products and other foodstuffs that tourists can easily consume during their visit. Other examples of food markets that underwent similar changes are Santa Caterina in Barcelona, San Fernando in Madrid and Porta Palazzo in Torino.[41]

The third type is the traditional food markets which are still focused on serving the local community. There are pressures for change as more and more tourists visit them. Their traditional nature is what makes them authentic, a characteristic that is appreciated by tourists. On the other hand, as more tourists visit them, they will adapt and change thus compromising the very authenticity that makes them attractive.

### 5.1.8 Popular culture and sports

For some attractions, it is popular or sports culture that creates the interest that motivates people to visit. It is not uncommon for sports fans to want to see the places where sports teams play. In the world of football, Old Trafford, Santiago Bernabéu Stadium and San Siro (officially Stadio Giuseppe Meazza) are included in the tourist guidebooks of Manchester, Madrid and Milano, respectively. Similarly, fans of media or music stars would wish to visit the places where their idols lived. In Salzburg, for example, the house where Mozart lived is one of the most popular visitor attractions in the city. The house where Mozart was born is also an attraction. Even films and literature can produce sights and locations that fans would want to visit. An early example of this is 221B Baker Street that readers of Sherlock Holmes books are so eager to visit. The house has been converted into a Sherlock Holmes museum. Another example is *Ulysses* by James Joyce – a book inspired by the city of Dublin. All the action of *Ulysses* takes place in Dublin on a single day (16 June 1904).[42] Some tourists seek out the sites associated with the book and there is also an Ulysses walk offered to visitors. Just because a story is fiction does not mean that the site has no meaning. The meaning of the place was created by the artist or author who was inspired by the real place. The fictional space then becomes real to the viewer or reader. Fans would want to visit the physical place that inspired the fictional work.[43]

In the case of films, fans would want to see the place and the setting where a famous film was shot. Some destinations promote themselves as ideal places for the shooting of films, making the film industry itself as a generator of tourism. One example is Malta which offers several settings for films at Valletta, Mdina and Cottonera. In these localities, many streets and urban spaces are untarnished by modern development, making them suitable for films that require a historic setting. The context combines with pleasant weather to make Malta an attractive option for film producers.

## 5.2 CATHEDRALS, CHURCHES AND RELIGIOUS BUILDINGS

In the modern age, European societies are becoming increasingly secular. Compared to, say, the latter part of the last century, there are significantly less people regularly practicing religion.[44] This notwithstanding, the built heritage associated with the Christian religious traditions retains an appeal that often transcends personal culture and beliefs. As a result of this interest, many cathedrals, churches, monasteries and other religious sites have effectively become yet one more element of the tourism product.[45]

There are three levels of tourism relevance of religious sites. At the highest level are the cathedrals, churches and religious sites which are of sufficient cultural and religious importance to motivate people to visit the city. Examples include St. Peter's in Rome, Notre-Dame Cathedral in Paris,[46]

Sagrada Familia in Barcelona and St. John's Cathedral in Valletta, Malta. Then there are cathedrals and churches which are of touristic relevance but not sufficiently to motivate tourists to visit the city. These would be part of the overall tourism product that is offered by the city, possibly also being a must-see attraction. One example is Wawel Cathedral, Krakow. Most churches in cities fall into a third category which are those churches that are not listed as a main attraction in the guide books but which are also interesting for religious, cultural, artistic or architectural reasons. A visit to the lesser-known churches and religious sites in a city enhances the overall experience of the visitor. For example, a visit to Salzburg would most likely include a visit to several churches located in or near the Old Town (Case Study VA3).

Some historic city centres seem to have a church around every corner. For example, Smørvik[47] notes that it is difficult to imagine Rome without its many churches. Sometimes they are small and unassuming, with a narrow church facade squeezed in between residential and commercial buildings. Other times they are massive buildings in prominent locations or dominating over a main city square.

Tourists who visit churches and religious sites do so for many reasons, some religious and some secular. Religious motive has a strong spiritual focus and involves the search for a deeper spiritual and fulfilling experience. Certain individuals choose sites linked to their faith or to a specific holy person or event that is relevant to their faith.[48] Although focused on religion, these visitors often have a broader interest in the historical and cultural dimension of the site.

People visit sites of religious significance for a variety of reasons that may or may not be related to faith or spiritual needs. For many visitors, a religious experience is not the purpose of the visit.[49,50] For example, in a study of visitors to four cathedrals in the United Kingdom, Winter and Gasson[51] found that almost two-thirds of visitors claimed to be motivated by historic and architectural interest. Less than a quarter of respondents visited for religious reasons. Similarly in a study of visitors to the Basilica of Santa Maria in Trastevere, Rome, only one of fifteen interviewees described themselves as religious.[52] Therefore, for a vast majority of visits to churches and religious sites, the experience cannot be described as religious tourism. Smørvik[53] describes religious tourism as a form of tourism 'whose participants are motivated in part or exclusively for religious reasons'. A visit to a religious site becomes religious tourism if there is a sense of pilgrimage and if the visit is motivated by religious belief.

Religious heritage sites offer multidimensional experiences that oscillate between the religious and the secular. Non-religious motivations can be a combination of factors. Some visitors to a church may be motivated by impulse or an urge that arises there and then, as they walk and explore the urban area. Curiosity compels exploring tourists to get to know the narratives and history of a place – narratives that until a few minutes earlier they might not have known existed. There is also a 'visual curiosity', with visitors wanting to set their eyes on features, including artworks that they had never seen before. As the tourist is exploring the historic centre, the open door of a church is inevitably seen as an invitation to walk in.

Smørvik[54] suggests other reasons for visiting churches and religious sites. Visitors may be seeking a break from the world outside or a peaceful experience. They may wish to enter the church because of its status or position. Hughes et al.[55] suggest that visitors to cathedrals may be looking for 'selfguided, contemplative and reflective individual experiences'. Even for non-believers, the visit to a church may be an opportunity to reflect and could therefore encompass something existential or spiritual. Smørvik[56] describes it as follows: 'The encounter with the interior of the church, the many lighted candles, the people praying at the front, the dark brown wooden benches, the sparse light from the high windows, and the low-pitched voices all invited visitors to rediscover their true self'. There are also heritage-focused visitors who are keen to learn about the cultural dimension of the place including its architecture and its associations with specific persons or events.[57]

Oriade and Cameron[58] explore the intricacies and tourism implications of cathedrals as one type of religious building. The cathedral is the principal church in a city and is usually the official seat of the bishop. They note that visits to cathedrals are becoming more popular even among non-religious visitors and hence they question whether any cathedral visit should be described as religious tourism. Cathedrals are often huge buildings and important landmarks in the city. They are normally a dominant feature in a city's skyline. For example, Cologne Cathedral dominates the city's skyline at a height of 158 metres. In some cities, the cathedral's skyline prominence has been diminished due to modern tall buildings in the central area of the city. The cathedral building often forms part of a series of historic buildings originally intended for religious purposes such as monastery, seminary and church administration offices. The cathedral quarter is normally a unique area offering remarkable townscape qualities making it an attractive place to visit. One example of this is the Ostrów Tumski district in Wrocław, Poland. Partly located on a small island on the Oder River, this area is the oldest part of the city. Apart from the Cathedral of St. John the Baptist, the area is occupied by other buildings that are still in use for religious purposes.

Oriade and Cameron[59] make many important observations about the tourism role of cathedrals. These are relevant also for other large churches that might not have the status of a cathedral but which have a central religious and cultural role in the life of the city. Although the central function of a cathedral is for religious worship, many tourism stakeholders and religious leaders are increasingly justifying cathedrals as a principal tourism resource, noting that cathedrals are the main reason why people visit the cities where they are located. Large historic buildings, such as cathedrals, are expensive to maintain and conserve. It is not surprising therefore that religious leaders see tourism as an opportunity for the long-term financial viability of the historic properties for which they are responsible.

Throughout the ages, the cathedral has had a central role not only in the religious life of the city but also in its cultural and social life. People used to – and still do – contribute large sums of money and this enabled faith communities in Europe, Catholics in particular, to build larger and more impressive religious buildings and to adorn them with the most notable works of art. In a modern-day society dominated by a desire for cultural consumption, it is inevitable that there will be a significant demand for visitation to a city's religious sites. The high demand for visitation offers a number of challenges to the managers. A question that cathedral authorities have to contend with is whether to charge an entrance fee or not. Some would argue that people should not be made to pay to enter a place of prayer as this would be incompatible with the cathedral's spiritual and social function. Most cathedrals do not charge an entrance fee although they expect visitors to make donations.[60] Another dilemma is that cathedrals were not meant for visitation by tourists and therefore, if not well managed, the presence of tourists in the cathedral may come in conflict with its function as a place of prayer. This is normally resolved by dedicating a small side chapel exclusively for prayer, allowing the rest of the cathedral to be appreciated by visitors.

Having a cathedral or church visited by tourists may have positive implications for the local community. It is an opportunity for them to show their religious heritage and, by implication, their way of life to a wider – potentially international – audience. It helps generate a sense of pride not only in their church but also in the city district area where they live.

### Case Study VA3: Churches in Salzburg Old Town

> In cities of a Catholic tradition, it is not uncommon to have several churches within or close to the historic centre. The author's own experience of Salzburg, for example, involved an extensive leisurely and exploratory walk of the city centre. Apart from many visitor attractions, the walk inevitably included visits to several churches. The better known churches were fully open to the public. Those that are less visited had the entrance area separated from the rest of the church by means of a wrought iron or glass partition allowing visitors to see the inside of the church without putting the church at risk of theft or vandalism. Within Salzburg Old Town (Altstadt), there are three churches and religious sites that are popular tourist attractions (Salzburg Cathedral, St. Peter's Abbey and St. Sebastian Church and Cemetery). There are

also churches which are less known to visitors but have remarkable charm and authenticity (Markuskirche, Kollegienkirche, Michaelskirche, Franziskanerkirche, Kajetanerkirche and Dreifaltigkeitskirche). These and other churches are all located within walking distance of each other in or close to the Old Town. Salzburg is a remarkable tourism destination and the possibility of entering and viewing so many churches makes a visit all the more remarkable.

## 5.3 NATURE-BASED ATTRACTIONS

### 5.3.1 Gardens and city parks

Gardens and city parks play an important role in the quality of life in towns and cities. Green spaces are valued for recreational, health, and environmental purposes, and for their contribution to city attractiveness.[61] Connell and Meyer[62] argue that garden visiting is a significant activity in cities and thus there needs to be knowledge and understanding of the kinds of experiences that visitors seek. They identify various ways that gardens provide enjoyment to visitors. There is the appreciation of the design and the ingenuity of assembly of soft and hard landscaping. In addition, there is the admiration of the attractiveness and beauty of plants individually, and also in combination with other plants. For visitors with no special interest in plants, there is the enjoyment of the scenic qualities offered by gardens as well as the simple enjoyment of being outdoors in a pleasant environment. In visiting gardens, it is possible to stroll quietly and enjoy nature as a release from stress. The enjoyment of gardens is reliant on the condition of the weather, the time of year and the time of day; all of which will have a greater or lesser impact on the visitor's perception of the garden. For example, autumn colours or spring flowers can change the appearance of a garden dramatically. The visitor experience is also shaped by several site-specific factors such as the range and quality of visitor services and facilities available, the ambience and the levels of crowding. The built facilities also affect the overall experience of the visitor and these normally include a tearoom or cafeteria, a shop, car parks, toilets, litter bins, a children's play area[63] and access ramps for people with mobility difficulties.

Some gardens have the added attraction of exhibiting extensive collections of plants, including ornamental plants. These are referred to as botanical gardens. Plants are labelled with common and scientific names and regions of origin. Beyond leisure, many botanical gardens have a scientific purpose; the study of plants. The number of plant species varies from a few hundred to several thousand, depending on the land area available and the financial means available. Some gardens dedicate a section to trees and shrubs. This is referred to as an arboretum. Botanical gardens in colder climates often have greenhouses to grow plants that would not survive the winter cold. The greenhouses are often used for display of tropical plants.[64] Some of them are managed and promoted as visitor attractions as is the case with the Tropical Houses in the botanical garden at Aarhus, Denmark. These are modern-designed greenhouses with numerous plant collections from four different climate zones. Another interesting example is the Sheffield Winter Gardens. This is an urban greenhouse with an indoor tropical garden. The difference here is the city centre location, providing a place of quiet refuge to shoppers and city workers, apart from being an attraction (Case Study VA4).

In larger cities, the richly landscaped gardens are often an adjunct to a palace or a stately home, designed and decorated in the style of the era in which they were built. From the sixteenth century onwards, right up to the nineteenth century, royal and privileged families across Europe built their lavish palaces and countryside retreats, often with extensive gardens and wooded areas, to symbolise their power and wealth. Many of these properties survive to this day virtually intact. Today the palace or the gardens or both are opened to visitors and they often become a main tourist attraction in their city. The best-known examples are Palace of Versailles in Paris, Schönbrunn Palace in Vienna and Nymphenburg Palace in Munich. Benfield[65] claims that gardens have a remarkable impact on tourism in some cities and cites the Royal Botanic Gardens at Kew as an example. Kew gardens received 1.8 million paying visitors in 2018 making it the third most visited paid attraction in England.[66]

Landscape architecture is a profession that is closely associated with gardens and city parks. It seeks to preserve and develop green open spaces in and around cities in a manner that achieves contact of city residents with the natural environment. A landscape architect considers the aesthetic and functional aspects in the design of the garden or park, both of which are essential to enhance the visitor's experience.

Similar to gardens, city parks offer enjoyment through closeness to the natural environment. City parks occupy a much larger area than what gardens normally would. There are variants but a typical city park would be mostly occupied by dense wooded areas with pathways making it possible for people to walk or jog through the trees, as is the case with Stockholm's National City Park (Case Study VA5). Compared to gardens, parks require less landscape design input since it is nature in its most pristine form that creates the attractiveness.

Referring to London's parks, Smith[67] describes the multifaceted ways in which parks are used as public spaces. He argues however that they are being commercialized because of pressures to generate income for their management and maintenance. Urban parks are being more intensively used as venues for commercial events. He laments the fact that every time a ticketed event is staged, the amount of genuinely public green space available to use is diminished. It also erodes London's reputation as a city punctured with green havens.

## Case Study VA4: Sheffield Winter Gardens

Sheffield Winter Gardens[68] is one of the largest temperate glasshouses to be built in the UK during the past hundred years. Inaugurated in 2003, it is an indoor tropical plants garden providing an important amenity for the public in the city centre. It creates a stunning green space and provides visitors with a unique experience. The urban greenhouse is home to more than two thousand five hundred trees and plants, from a wide diversity of one hundred and fifty different species. Most are from the world's southern hemisphere. They are strikingly exotic and unusual in their forms, textures and colours, making them radically different from Britain's own native flora. The greenhouse structure has a length of seventy metres. It consists of a series of huge timber arches, twenty-one metres high, with glazing in between. The timber is Glulam which is made by forming and gluing strips of timber into specific shapes. The Gardens were part of a wider regeneration project for Sheffield that included a remodeling of the city centre's urban spaces. This is discussed in Case Study CC2 in chapter 2. The building won several awards and is considered to be an important iconic feature in Sheffield's city centre (figure 5.3).

## Case Study VA5: Stockholm's National City Park

In 1995, the Swedish Parliament designated an area of approximately twenty-seven square kilometres as National City Park to protect cultural and natural heritage in metropolitan Stockholm.[69] The area spans the municipalities of Solna, Stockholm and Lidingö and includes the royal park of Djurgarden. The Swedish Environmental Code enacted by Parliament states that: 'The Ulriksdal–Haga–Brunnsviken–Djurgården area is a national urban park. New development, new buildings and other measures shall only be permissible in national urban parks if they can be undertaken without encroaching on park landscapes or the natural environment and without detriment to any other natural and cultural assets of the historical landscape'.

Previously in the area, various proposals for transport infrastructure and housing projects were a source of concern for environmental NGOs keen to protect Stockholm's urban greenery. Eventually in 1992, these came together under a single umbrella group, the Eco Park Association, to lobby for the formal designation of the park.

Figure 5.3  **Sheffield Winter Gardens.** *Source:* Author's own.

Designation was motivated by the area's recreational, ecological and national cultural value, as well as its relevance to the nearby urban areas and to the city. The designation process was not without tension. For example, the ideas of pristine nature and nature conservation clash with the possibility of people engaging in more formal forms of recreation such as organised sports. An initial difficulty was to define the boundary of the park since the area already contained various buildings and structures such as the modernistic main building of Stockholm University, museums and major roads.

Upon designation, overall coordination for the area was entrusted to a fifteen-member National City Park Council that included representatives of the many stakeholders with an interest in the area. Agencies responsible for managing various parts of the area were represented, namely the municipalities of Solna, Lidingö and Stockholm and the Royal Djurgarden administrative authority. Also represented were the Stockholm University, the World Wildlife Fund, museums and the Eco Park Association.

People enter the park for recreation, natural and cultural experiences, walking, biking, swimming, sports, and picnics. The park is a place where one can experience rare species, unspoilt nature, serenity and silence. It is also a place for various leisure activities such as visits to museums, theatres, restaurants, cafes, and the amusement park.

The area has long been valued for its beauty and rich biodiversity. The park's landscape serves to maintain biodiversity, both by providing green corridors and by hosting valuable animal, bird and plant species. Specific sites within the park are replete with deciduous trees. There are also several interesting geological features. The recreational aspect of the park is mostly focused on the experiencing of nature but it also refers to sports activities and other forms of leisure. The latter are concentrated in the southern part of the area and include an open-air museum, an amusement park, theatres and restaurants.

In tourism information on Stockholm, the park is featured as part of an overall narrative that depicts the city as a place that offers the possibility of wholesome activities and recreation in beautiful settings.[70] The urban characteristics of the city are tacitly indicated by drawing on contrasts between nature and the city. Tourism information associates nature and experiences of it with peacefulness, beauty, and freedom from the stress of everyday life. The establishment of the National City Park was part of an ongoing process of 'place construction' or place making through effective coordination and communication, as well as better management – a process that continues to the present day.

### 5.3.2 Zoological parks and aquariums

Zoos are a form of museum but, unlike other museums, their exhibits are living creatures. In this regard, zoos may be considered as a substitute for the 'real experience' of visiting animals in the wild but that necessitates placing them in captivity.[71] They have a broad range of scientific roles such as observation and classification as well as the study of genes, mating behaviour, animal physiological and veterinary science. Since the latter part of the twentieth century, there has been growing concern about the welfare of animals held in captivity given that their natural habitat is the wild. Environmental campaigners argue that captivity is cruel to animals and that there is an incompatibility between the claimed educational and conservation roles of zoos.

It is not just zoos that exhibit wildlife. To classify the range of wildlife attractions, Shackely[72] used the concept of 'mobility restriction' and created a scale from 'complete confinement' at one end to 'complete freedom' at the other. She considers safari parks and nature reserves to be located near the 'complete freedom' end of this scale, whereas aquaria and butterfly parks[73] are near the 'restricted' scale end.

There are trade-offs in the tourist experience associated with each scenario. In a zoo, the visitor is virtually guaranteed to see most zoo animals at a relatively close distance, but there is minimal habitat context and no thrill in discovery. In a wildlife park, the animals are allowed much greater freedom over an extensive area of natural environment. The downside is that the animals are seen at a greater distance and at certain times might actually not be visible to visitors at all. In some cities such as Berlin, Copenhagen and Rotterdam, zoos are popular attractions with visitor numbers comparable to some other important attractions in the city.[74] Despite their significance as visitor attractions, the popularity of zoos has declined since the 1990s, partly because of the rise of competing attractions. Moreover, the fact that the animals are in captivity is a deterrent for some people to visit.

Aquariums are another type of visitor attraction that are popular in many cities. Across Europe, there are several interesting and unique aquariums to visit. Based on the size of tanks and the number of big fish, the three best aquariums in Europe are Oceanogràfic in Valencia (Spain), the Nausicaá in Boulogne-sur-Mer (France) and Oceanário de Lisboa (Portugal).[75]

## 5.4 THEME AND AMUSEMENT PARKS

In many cities, purpose-built leisure parks are major attractions for leisure and tourism. Leisure parks for entertaining the public have a long history in Europe. Open since 1583, Bakken near Copenhagen claims to be the oldest amusement park in the world. By the nineteenth century, entertainment parks were firmly established. The most renowned were the Tivoli Gardens in Copenhagen and the Blackpool Pleasure Beach, opening in 1843 and 1896, respectively.[76] Both are still highly visited parks with millions of visitors annually. The distinction between a theme park and an amusement park is not always clear although the former is loosely based on some theme, whether geographical, historical or other concept.

To be successful, an amusement park requires effective management. Crowd management is required to ensure that no part of the park is subject to excessive crowding and that the queues for the rides are kept short to minimise waiting times. Maintenance of equipment is a must to ensure that all equipment works efficiently. Poor maintenance of mechanical equipment could lead to incidents in which people are hurt. Buildings and structures have to be kept well maintained and cleaned, including periodic painting. An amusement park is dynamic in that new rides are created to replace older ones to ensure that the rides on offer meet the demand and expectations of clients. This requires heavy investment. For example, Blackpool Pleasure Beach designed and constructed more than twenty rides over a period of fifty to twenty years to replace rides which had become obsolete. Moreover, in 2011, the theme park invested 10 million sterling to develop Nickelodeon Land, a five-hectare theme park for children within the main park. This replaced the park's previous children's area.

Amusement parks can be classified according to size into three categories.[77] The first is local parks that largely cater for the day-tripper market. Then there are flagship attractions that draw national markets and a significant number of foreign visitors. Tivoli Gardens in Copenhagen and the Prater in Vienna are two examples. The third category consists of iconic parks occupying large areas of land. These have become destinations in their own right and attract a worldwide market. The most visited theme park in 2019 was Disneyland Park in Paris with almost 10 million visits. Other popular parks include Europa-Park in Germany and De Efteling in the Netherlands each with around 5.5 million visits in 2019.[78] There are numerous other theme parks across Europe – most of them are in cities or within easy travel distance of one or more major cities.[79] The greater the number of residents within the catchment area of the theme park, the greater its commercial viability. Proximity to an airport is an added advantage as that enables the park to attract an international market.

A water park is another form of entertainment facility. The main one in Europe is the Therme Erding in Munich with 1.9 million visits in 2019. Other water parks worthy of note are Tropical Islands near Berlin and Aquapalace Praha in the Czech Republic both with around 1.2 million visits annually.[80]

On a much smaller scale, there are innumerable other forms of tourist entertainment like amusement arcades, escape rooms, model railways, waxworks (including increasingly popular 'dungeon' settings) and similar small commercial exhibitions. In smaller city destinations, they have an important role and therefore it would be wrong to ignore their relevance within the organisational structure of the attractions industry, even though for many, income will be relatively small. For these smallest types of visitor attractions, their very existence is tenuous – often run as a family business or with the help of voluntary workers.[81]

*Chapter 6*

# City Resources for Tourism (3)
## Accommodation and Other Facilities

In a previous chapter, a distinction is made between the core tourism product and the supporting facilities and services (the primary elements and secondary elements, respectively). The latter are vital for a city destination to be successful but very rarely are they the reason for the visit. The most important supporting services are the tourism accommodation but other facilities are also very relevant and need careful consideration.

## 6.1 TOURISM ACCOMMODATION

### 6.1.1 Hotels

Hotels have an important role in cities. They provide the opportunity for visitors to stay for a length of time, be it for business or leisure. Tourists require a location where they can rest and stay overnight when they stay at a destination that is not their home. Hotels are generally the largest item of tourist expenditure in the city[1] making the hotel sector the largest and most consistently present sub-sector within the tourism economy. Hotels generate economic activity that contributes to the local economy. Tourists buy the service of accommodation and, in turn, the hotel buys goods and services, mostly from the same city.

Any city that aspires to be an important tourism destination will require a good supply and choice of hotels. The growth of tourism activity in a city is normally accompanied by a comparable growth in the provision of tourism accommodation. There is a great diversity in hotel accommodation provided at most city destinations. The size of the hotel establishment can range from a small bed and breakfast operation with just a few rooms to extensive hotel blocks or complexes with a capacity to cater for several thousand guests. There is also great diversity in quality ranging from the basic, highly functional form to extreme luxury and opulence. The ownership and management ranges from the private and informal, normal for the smaller establishments, to the large hotels operated by large multinational commercial organisations.[2] There is also diversity in the way the hotel product is marketed to customers. To secure business, large luxury hotels emphasise facilities and quality service to certain market segments, such as business travellers. In contrast, budget hotels emphasise price in their promotion. As illustrated in figure 6.1, these factors impact the way the accommodation product is constructed, portrayed and sold to customers.

A hotel is not simply a property offering accommodation and catering. It is a business oriented towards a constantly changing clientele.[3] The traditional view of a hotel was an establishment providing accommodation as well as food and beverage services to short-stay guests against payment. This is changing with constant innovation, evolution and diversification in the accommodation. Larger hotels are providing leisure, sporting and entertainment facilities as well as business and conference services.[4] The accommodation product is a complex amalgam of factors that combine to provide the tourist with something they wish to consume.

Tourism accommodation assumes many forms, and not all of them fit the conventional image of the hotel. The accommodation sector can be divided into serviced accommodation, such as hotels, and non-serviced

Figure 6.1 **Tourist accommodation as a product.** *Source:* Adapted from Page, 2011.

accommodation such as self-catering apartments. Each sector develops in response to different markets.

In the 1950s and 1960s, hotel brands and chains began to emerge with the focus being on standardisation of products and services.[5] This trend was best epitomised by the famous Holiday Inn slogan in the mid-seventies: 'The best surprise is no surprise'.[6] The slogan referred to the reliability of the hotel chain's service and quality.

The location of hotels is an important topic to understand the dynamics of tourism in cities. The traditional location for urban hotels was the city centre with many nineteenth-century hotels being built near railway stations. In the twentieth century, rail travel remained an important mode of transport and therefore a common feature in most cities is a concentration of hotels in the area around the main train station. As the hotel industry developed, close to train stations remained the best place to locate hotels to better serve their customers. It was the place to go to immediately upon arriving at a city, before going on to do business in the city. Before leaving the city, visitors are likely to collect their bags from the hotel before taking the train.[7]

The location of a hotel in many ways defines its market and its character. City centre hotels normally cater for business traveller, and hence the location near major financial institutions. The city centre almost always includes luxury hotels as they can afford to pay for the most accessible and prestigious sites, even if they have to compete with other uses. As cities changed and developed, the demand for hotel accommodation grew in areas of the city other than near the train station. Some hotels were located just outside the business and retail zones, away from the highest land values. Others were built at the fringes of the city, thus reinforcing the idea that hotels occupy 'gateway' locations.[8] Such hotels are most suited for visitors arriving by car who would then use public transport to visit the city's visitor attractions. Another cluster of hotels is often found close to city airports, catering for the needs of departing and arriving travellers and also for business conferences attended by executives flying from distant places. In London, many hotels are located on the west side close to the roads that link London's central areas to Heathrow airport. Other markets that drive location are hospitals, government centres, convention centres, law courts, universities and city authority offices.

In cities with good leisure facilities (visitor attractions, museums, etc.), hotels cater for both business travellers and leisure tourists. Hotels provide a base for conferences, meetings and other forms of business travel. These are lucrative as business travellers staying in hotel accommodation have a higher propensity to spend while they are away than when they are at home. Hotels not only meet the visitor's basic requirement of shelter for the night but also add value to the experience by providing ancillary services. Hotels also have the advantage that hosting guests has the potential to generate additional revenue from food and beverage services.[9]

After long periods of neglect, many cities have turned to their waterfront for regeneration and new development. The regeneration of waterfronts often include hotels, capitalising on the air, light and open views afforded by the open expanse at the water's edge. Views add real value to a hotel property. Hotels demand higher rates for rooms with views of the skyline, of waterways or of green parks, than for those facing an adjacent office building. Higher rates can also be achieved for location even if without good views such as, hotels within walking distance of an attractive park.

At city centres, there are good transport links (metro, trams) and the use of private vehicles is more expensive (because of congestion charges, high parking fees, etc.). Hotel clients are more likely to rely on public transport or walking to move about the city. Many of the city's restaurants, museums, shopping and office buildings are usually within walking distance. Business travellers will choose a hotel that is within walking distance of the office/s where they will have their meetings. City centre hotels therefore do not normally provide parking for their residents or users of their other facilities. In parts of the city where the development is less dense or where public transport links are not as good, there may be a case for hotels to provide some parking. Further out, at the periphery of the city, the provision of some parking for hotel residents becomes a necessity.

For an investor where to locate a new hotel is an important decision – a better location will result in increased occupancy and therefore increased profits. Luxury hotels are usually able to pay for the most accessible and prestigious sites, although they must compete with other land uses. Budget hotels are often pushed to sites on the edge of the city centre where land values are lower. When hotels are built to serve specific markets, they will locate near to the optimum site. For example, a major conference centre requires tourist accommodation to maximise on its potential and hence new hotels will be located nearby. Transport infrastructure is another important consideration for the location of new hotels. For example, if there is a major tourist attraction along a well-established bus or tram route, the investor could consider locating a new hotel close to a nearby stop.

There are several variants of the hotel accommodation model that merit some attention namely hostels, diffuse hotels and pod hotels. Hostels are much cheaper than hotels with guests being accommodated in dormitories with shared canteen and sanitary facilities. Hostels are a key component of the youth travel accommodation mix. Their unique selling proposition is the offer of a social experience as well as cheap accommodation. Whether with locals or other travellers, social interaction is an important motivation for young travelers.[10] Diffuse hotel is another variation of the hotel model.[11] The idea was developed in Northern Italy in an attempt to give new life to smaller villages that were being abandoned. It makes use of buildings in backward regions to provide accommodation to tourists who wish to experience the community life in towns and villages that are not normally frequented by tourists. The accommodation is provided in separate houses within walking distance of each other with reception services to tourists being provided from a nearby office. The difference between the diffuse hotel and short-term tourism accommodation has become increasingly blurred as units can be sold in either manner.

A radically different model is the capsule or pod hotels. This was developed in Japan as a way for budget travellers to have somewhere safe and affordable to sleep. The accommodation is suitable in megacities where the cost of city centre accommodation is high and where the time and cost needed to travel from the outskirts to the city centre are substantial. The primary market for capsule hotels is young professionals who live in the outer suburbs of a megacity. Often kept late at work and dreading the long-ride home, they frequently seek inexpensive overnight lodging. These hotels typically contain hundreds of units or capsules. A capsule is a sleeping compartment of approximately two metres long by one metre wide and one metre high. They are stacked two high and double-loaded on long corridors.[12] The public areas of the hotel include reception, a cafeteria and other facilities. Although common in Asia, it seems that the capsule hotel model has not caught on in Europe.

### 6.1.2 Boutique hotels

The hospitality sector has been characterised by decades of brand standardisation in which chains such as Holiday Inn, Marriott and Hilton provided guests with consistency in the hotel product. The boutique hotel concept was developed in the early 1980s as a reaction to the standardisation of hotels. Guests of mid- to upper-income levels have been looking for hospitality experiences that are different from those offered by the large hotel chains.[13] The first renowned boutique hotels to open their doors to the public were the Blakes Hotel in South Kensington, London, and Morgans Hotel in New York.[14,15]

With an increasing number of boutique hotels in cities across Europe, this hotel type has become an important accommodation component for business and leisure travellers. Guests nowadays expect more than simply comfort and convenience. Since the 1990s, more tourists are creating their own personalised holidays-seeking accommodation in boutique hotels and looking for innovative experiences in unique settings and superior quality. Customers' expectations are shifting towards wanting an experience from the accommodation and not just a bed for the night. An increasing number of guests prefer to be 'surprised' in a positive way. When planning trips, they seek properties that are noticeably different in look and feel from the conventional chain hotels. Some may choose to stay in this hotel type because it is fashionable to do so. In this regard, boutique hotels are marketed for the experience and the image they conjure up for prospective customers. These customers normally have the benefit of extensive travel experience with an extensive range of online information and products to compare with. Unfortunately, there are thousands of hotels across Europe that are promoted as boutique hotels when in fact they are not. This misrepresentation has somewhat tarnished the image and perceived value of boutique hotels.

In French, the word 'boutique' refers to a small retail outlet that specialises in elite and fashionable items. The emphasis is on small size and specialisation. Despite the considerable interest in this segment of the hospitality sector, there is a lack of consensus of the terms 'boutique hotel'.[16] Despite the lack of a formal definition, there are features that are widely considered to be characteristic of a boutique hotel.

Their differentiation from chain hotels is derived from the design, artistic, cultural or historical appeal, aligned with the prestige and exclusivity of the property.[17] Boutique hotels integrate functional and aesthetic elements such as style, layout and architecture. In terms of design, there are two main elements to consider; the building itself and the internal design features. The architecture has a vital role to play but what is even more important for the boutique hotel experience are the internal design features. The combination of design and service is what makes each boutique hotel unique. 'Experience' is a key theme in describing boutique hotels.[18] Another characteristic to describe a boutique hotel is the limited number of rooms, normally fewer than 100 rooms.

The success ingredients of boutique hotels are uniqueness, good value for money and a superior product accompanied by an equivalent superior service. Boutique hotels are expected to be charming, intimate, luxurious, cutting-edge, avant-garde, trendy and uniquely designed with attention to detail, providing a high level of personalised service. In cities with significant historic areas, boutique hotels are often associated with heritage buildings such as small palaces and historic homes. The downside of historic buildings are the high restoration and recurrent maintenance costs although these are offset by room rates which are higher than standard hotels.

The ultimate point of differentiation of boutique hotels is the need to avoid the uniformity of corporate hotels. That said, several larger chains are beginning to include boutique style hotels in their portfolios as they begin to recognise their market appeal, especially among discerning travellers.[19]

Boutique hotels is a hotel type that falls into a wider generic categorisation namely that of design hotels. The salient feature of design hotels is the dedication to immersing the guest in a world of exquisite and interesting design featured both in architectural innovations and unique interiors. Design hotels tend to focus on innovative, experimental or unconventional design. They can position themselves in the market anywhere from budget to luxury. Over the past few decades, there has been a sharp increase in the popularity of design hotels, with hundreds of new properties claiming to be a design hotel. There is a high correlation between properties that focus on architectural design and their profitability. Investment in sophisticated architecture and design pays off, even if operating costs often exceed industry standards.[20] This trend has spread with incredibly interesting and unique design showing up at airport hotels, mixed-use projects, office-park hotels, ski resorts, casino hotels, marina hotels and so on.

### 6.1.3 Short-term rental tourism accommodation

The traditional market for tourism accommodation involves renting rooms from hotels or other similar types of business operations. Airbnb has shaken up that model by enabling an ordinary person to rent accommodation from another ordinary person using an online marketplace (sometimes referred to as 'peer-to-peer accommodation'). Short-term rental tourism accommodation (henceforth referred to as STRs) is common in many cities especially those that are established tourism destinations. While not a new phenomenon in cities, it has grown exponentially following the advent of digital platform companies. Airbnb is the best-known platform and the one that is most commonly used for this type of accommodation but there are several others. The widespread use of internet technology gave potential tourists access to information on accommodation and travel without the need for a commercial intermediary. It also made it possible for individuals to make their own bookings. Moreover, STRs for tourists significantly increased the choice and availability of tourism accommodation. Travel became more affordable making it possible for people to take more frequent holidays. The growth of STRs as a form of tourism accommodation was one of several factors that brought about significant changes in the way tourism operates. Another important change was the growth of low-cost airlines that brought with it lower fares and increased connectivity between cities. These changes brought about increased tourist flows, not only to major cities and the traditional tourist hotspots but also to medium-sized and secondary cities that now saw new opportunities for expanding their tourism economies.

Owners of residential property can effortlessly promote their STR accommodation to potential guests using digital platforms that link owners with potential guests. Airbnb has made it simple for hosts to post information about their properties including photographs. They can also take reservations and receive payments. The Airbnb platform enables the establishment of trust between host and guest. This is achieved by having hosts and guests posting online reviews about each other. Over time a succession of positive reviews enable the hosts and guests to develop a profile of reliability and integrity about themselves. Trust is also developed by means of direct communication between host and guest.

The highest densities of STRs are in city centres and near major attractions although they are also present across the city including in areas that are predominantly residential and lacking tourist attractions. In residential areas the short rental guest will live like a local and experience the day-to-day life of the area. This includes interacting with the host and using the same local shops that other area residents use. Compared to staying in a hotel in the city centre, staying in Airbnb accommodation provides a more authentic experience of the city. The traveller can 'stop being a tourist' and actively engage in the ordinary urban life of the city.[21,22]

STRs are beneficial to city destinations as they help foster tourism. An obstacle to travel is cost as low- and medium-income earners consider carefully whether they can afford the long-awaited holiday at the destination of their choice. The availability of more affordable accommodation greatly mitigates this hurdle and makes it easier and less costly for people to travel.

Renting to tourists provides advantages to property owners. STRs can be priced very competitively because there are generally minimal labour costs and, for rooms in owner-occupied properties, the cost of rent and electricity are already covered. Besides, the hosts are not financially dependent on the revenue. In some countries, it is relatively easy for the host to evade paying taxes on the income from the STR.[23] The additional income generated from STRs enables first-time homebuyers to partly cover major expenses such as the repayment of house loans, thus reducing the possibility of repossession. It can also enable people with low incomes to cover basic living expenses.

STRs also generate benefits to the city. The tourist spend is more likely to be spread across the city in neighbourhoods that typically do not receive much tourist expenditure. In jurisdictions that legalise STRs, these can be taxed and thus generate revenue to city authorities. Some destinations suffer from significant differences in visitor numbers between peak and low seasons. Such destinations are able to deal with the spiked demand in the peak season by means of rooms provided in the STR market. Similarly, a city wishing to host a major event will have less concerns knowing that the increased demand for accommodation during the event can be catered for by STRs.

The expansion of STRs in cities is having a disruptive effect at various levels. This new business model poses a major challenge to the conventional accommodation industry. Compared to hotel accommodation, they are lacking in many features and requirements that are important to tourists, including service quality, staff friendliness, brand reputation and security.[24] At the destination level, it has generated several negative externalities. Critics claim that it increases rental and living costs in ordinary neighbourhoods.[25] Füller and Michel[26] argue, however, that these effects are the cause of much broader housing market dynamics with STRs simply contributing to a limited degree.

## 6.2 FOOD AND DRINK ESTABLISHMENTS

Food and drink are obvious requirements of a visitor to a city. Although eating is evidently an essential need, it may also be a pleasurable experience in itself. A city that aspires to be successful in tourism should have a wide array of catering facilities, especially in those areas where tourists are mostly present.

In recent years, there has been an increase in the demand for restaurants. This is understood to be any establishment that serves food and drink that can be consumed on the premises.[27] Other forms of catering facilities include cafes, bars, fast-food outlets and takeaways. There are several socio-economic factors that have brought about an increased demand for restaurants. These are mostly related to greater affluence and changing lifestyles. More people are travelling away from home whether for work, education, to shop or for leisure purposes and while away will need food and drink. With increased full-time employment, less time is available for people to prepare meals at home. There are more one-person households including young people or persons from broken families. A further reason for the increased demand is that more people are eating out as a leisure activity.

In most cities, the demand from residents for food and drink is high so only a relatively small proportion of catering facilities are patronised mostly by tourists. Thus, the locational distribution of catering facilities across the city is determined mostly by the demands of residents, office workers, students and others, rather than by the demands of tourists.[28] Having said that, there will be a tendency for catering facilities to cluster in areas where there are high number of tourists. Food and drink establishments are normally located along streets where there are many pedestrians passing by. Increased visibility makes them more competitive and commercially viable. The typical pedestrian street in any city centre would have street frontages of shops as well as food and drink establishment. A range of food and drink establishments are normally available at city centres and other areas where people congregate. The variety of cuisine ranges from the ethnic (e.g. Chinese, Indian or Italian cuisines) to the traditional; from gourmet cuisine to the menu of fast-food chains (e.g. burgers and chicken). The clustering of food and drink establishments in an area benefits tourists, as well as city residents. Customers often go to an area where there are restaurants to choose from and then take a last-minute decision where to dine. Clustering of catering establishments offers choice.

Often establishments offering food and drink combine with entertainment facilities such as nightclubs, discos and casinos.

C. Law[29] classifies eating out into two types namely 'body food' and 'soul food'. Body food is consumed to keep oneself going. It may be cheap, simple and eaten relatively quickly. It is the kind of meal taken during a work break, in the middle of a shopping trip or prior to engaging in a leisure activity. Soul food also satisfies basic needs but more than that it is sought after as an enjoyable experience. The food will be more sophisticated, the surroundings more pleasant and the expense greater. Longer time is taken to consume the food. It is also a social event rarely undertaken alone and often as a couple, family or group of friends. It may also be the type of meal offered as part of an important business encounter. In a tourism context, food is increasingly becoming an attraction in its own right. The consumption of local food is, in a sense, a means by which the tourists 'consume the destination'. A memorable food experience will greatly enhance the tourist's overall experience of the destination.

Depending on the location within the city, the demand for restaurant facilities varies throughout the day and throughout the week. Restaurants in city centres are normally busy at lunchtime and also very often in the evening. At lunchtime customers are more likely to be shoppers and persons working in nearby offices. If the area is pleasant or if there is a main visitor attraction nearby, many of the customers at lunchtime are likely to be tourists. In the evening, diners are likely to be people on their way to the theatre or to a concert or people wishing to have an enjoyable night out with friends or family.

## 6.3 FACILITIES FOR SHOPPING

Tourism is an important leisure activity. Shopping is also a leisure activity and therefore there is an affinity between the two, reflected by the fact that shopping is an activity to which many tourists dedicate some time during their holiday.[30] Even if tourism and retailing are not normally associated, shopping is an important part of any tourist's activities with tourists spending a significant amount of time and money on shopping.[31]

For many, shopping is a pleasurable experience and something they like to do when they have the time, as they do on holidays, and are not under pressure to do domestic chores. In leisure shopping, people derive pleasure from the aesthetic and experiential activities of browsing through commodities. It is an activity devoted principally to shopping for clothes and accessories.[32] Shopping is made more enjoyable with a pleasant environment in the shops and also in their immediate surroundings. Cleanliness, attractive shop fronts and provision of street furniture are all important to tourists. In addition, a lively atmosphere in the shopping street makes shopping more enjoyable and could therefore increase sales. As discussed in chapter 7, this is one of many reasons why having pedestrian streets in the city centre is a useful objective for any city that desires to enhance its competitiveness. Retail outlets and complexes redouble their efforts to compete in offering consumers bigger, better, more exciting and more memorable experiences than ever before.[33]

According to D. Timothy,[34] the relationship between shopping and tourism can be seen in two primary forms: shopping tourism and tourist shopping. Shopping tourism entails people traveling specifically to shop, normally to destinations that promote shopping as one of many leisure activities. The tourist's selection of the shopping destination is based upon the shopping attractiveness (available products and price competitiveness), the destination's tourist appeal and also the destination's reputation for shopping. Tourist shopping refers to tourists engaging in shopping as an ancillary leisure activity, often to fill time or to acquire souvenirs. Often tourists stumble upon retail opportunities or they seek out shops once in the destination. The distinction between shopping tourism and tourist shopping is important as each has its own set of motives and experiences. For the former, retail is the main motive of travel, so satisfaction with the trip is dependent on factors related to the shopping activity (the attractiveness of the retail complex, product availability and service quality) and the act of shopping itself. These tourists are less likely to be satisfied with the trip if retail conditions do not meet their expectations. For tourist shopping, on the other hand, shopping is a secondary activity and less likely to affect the holiday experience negatively as the act of shopping may be an insignificant part of the overall tourism experience.[35] While some tourists might be intent on purchase, others are casually browsing and window shopping. They do so as an adjunct to other activities for less tangible reasons such as fun, interest, sensory stimulation and social interaction.[36]

Within or in close proximity to a visitor attraction, one is almost certain to find shops, normally souvenir and gift shops. Historic attractions are often supplemented or even engulfed by retail commercial outlets.[37] Cities with heavily trodden tourist routes end up with the whole road corridor lined with shops. One example is the roads leading to Ponte del Rialto, Venice. Another example is Prague's Royal Mile which links the Old Market Square to the Castle via Charles Bridge. The outlets which are found in these areas vary from simple gift and souvenir shop to outlets selling speciality goods. An interesting combination is a historic bridge that at the same time fulfils the function of a shopping hub as is the case at Ponte Vecchio in Florence, Italy. The medieval bridge still has shops built along it, as was once common. Occupied initially by greengrocers and butchers, the shops today sell jewellery, artworks, gifts and souvenirs. At the same time, the bridge is one of Florence's main tourism attractions.[38]

Historic environments have come to provide an ideal setting for leisure shopping. The reuse of buildings for tourism-related retail provides new life for otherwise redundant buildings. The success of Covent Garden in London is a good example of how the conservation of a redundant market hall in a pleasant pedestrianised shopping district can become a tourist attraction and part of London's tourism product offer.[39]

In towns that are popular with tourists, retail provides the biggest opportunity for financial gain for property owners. Historic cities such as Bath, Chester and York have retailing centres which are far larger than the size of the resident population would suggest.[40] The downside is that retail servicing local residents are pushed out to be replaced with souvenir shops, smart cafes and specialist retail outlets of little use to local residents. A careful balance is needed between creating attractive retail facilities for tourism and leisure visitors and concurrently ensuring that the retail needs of local residents are provided for.[41]

## 6.4 VENUES FOR CONFERENCES AND EXHIBITIONS

The conference and exhibition industries generate significant economic activity and many cities have actively taken measures to tap into these market.[42] Many cities across Europe possess purpose-built venues for the staging of conferences or exhibitions. Most are owned by the city authorities, with their development or extension being financed in part through state or EU grants. In most cases, a conference or exhibition centre is run by a public-private partnership, with the city authority being the majority stakeholder. Often losses are incurred on the operating costs and these are borne by the taxpayer.[43] The argument for public funding support for such venues is based on the economic benefits that the staging of major conferences and exhibitions can bring to the city, including the attraction of more visitors, increased tourist expenditure and higher occupancy rates in hotels. Another frequent motive for the public support of conference or exhibition venues is for the urban regeneration of an area of the city.

### 6.4.1 Venues for conferences

Conferences are sometimes held in buildings and facilities that are intended for other purposes such as museums, university campuses and public sector offices. Many larger hotels also provide excellent conference facilities as this is often seen as a means for increasing occupancy at times of the year when the demand from leisure tourists is lowest. To compete effectively in this market, however, a city requires purpose-built conferences centres.

Purpose-built conference centres are often provided with striking architecture and imposing exteriors. They seek to portray a strong and memorable image to visiting delegates as well as to the city's own residents. More importantly they provide extensive internal spaces. This is required to accommodate the largest types of conferences, such as annual meetings of international associations, where the number of those attending can run to several thousands. These are typically well equipped with the latest audio-visual technology and booths for simultaneous translation, as well as numerous ancillary spaces for breakout rooms or for catering.

Many venues are intended not only for conferences but also for the staging of entertainment, performances and cultural events for city residents and leisure visitors. Many auditoriums are adaptable for concerts and for theatrical performances. Flat-floored halls are invariably equipped to be used for meetings, exhibitions or functions (receptions, dinners and banquets). In some cases, tiered seating can be removed to leave a flat floor for alternative requirements. Multipurpose use of a venue enhances its financial viability as it increases the number of days in a year when it will be in use, even if providing for adaptability in use will increase capital costs.

Planning for a major purpose-built conference centre can be challenging because it is difficult to anticipate future demand. Like for any major urban facility, the lead time from the initial idea for a conference centre until its opening can be as much as ten years. The process involves among others identification of a suitable site, design and planning, establishing financing mechanisms, construction of the venue and related infrastructure.[44] Moreover, a management structure for the venue has to be established, including recruitment and training of staff. In such a period, substantial changes can happen in the conference and related industries. The disruption and uncertainties caused by the pandemic in 2020 is a case in point.

University campuses possess a wide range of learning spaces, such as auditoria, lecture theatres and classrooms. These are highly suited for conference events and therefore it is no surprise that many universities seek to host conferences to earn additional income. Many universities have dedicated conference business units, separate from other academic activities, to run their conference business. They also offer residential conferences making use of the students' campus accommodation, even if these may be much less than five-star standard. For conference organisers and delegates, the advantages of staging a conference at a university campus often include good value for money, pleasant campus environment and location in an attractive area of the city.[45] Apart from purpose-built facilities, hotels and university campuses, conferences are sometimes hosted in what Rogers[46] refers to as 'unusual venues'. These are buildings whose main purpose is a use other than conferences such as museums, historic houses, monasteries, art galleries, visitor attractions and sports complexes. The attraction of unusual venues is that it gives the event a special appeal and can make it memorable for years afterwards, especially for delegates weary of yet another conference held in a conference centre or hotel with little attraction.[47]

### 6.4.2 Venues for exhibitions

Exhibitions are places that facilitate the exchange of ideas and information between exhibitors, industry specialists and visitors. Other words sometimes used for 'exhibitions' are 'expositions' and 'trade fairs'. There are two categories of exhibitions: business to business and business to customer. The former is usually restricted to those seeking to purchase products or services for use in their business or profession. Specialised trade fairs generally attract visitors and exhibitors from across the country where the exhibition is located and also from overseas. Such exhibitions stimulate significant domestic and international tourism to the host city and country. The latter are consumer fairs that are open to the public and feature any product that people are prepared to purchase like, for example, household goods and appliances, cars and financial products. These attract large number of visitors mostly drawn from the city itself and from the region. Most attend for a single day. Many larger exhibitions are intended for both the trade as well as the general public. Normally the first day or two are dedicated to the trade and the press, with the remaining days being open to the public.

For the companies who exhibit, the main purpose is to generate sales and also to promote new products. Exhibitions are a key component in their communications and marketing mix.[48] It is a cost-effective way for companies to communicate information to potential buyers, and also to receive immediate feedback on new products and services. Beyond actual sales and generating contacts, exhibitions increase the visibility and enhance the image of a participating company.

The advantages offered to visitors are also many. Exhibitions bring together under one roof an extensive range of goods and services. Products can be seen, handled and compared by potential buyers. Some products can even be smelled and tasted. They also offer visitors the opportunity to have face-to-face discussions with those who are knowledgeable about the product or service. Experts will provide them with authoritative answers to their technical questions. Many exhibitions include specialised lectures and conferences on aspects related to the subject and products being exhibited.

Large exhibitions generate considerable economic benefits to the cities where they take place largely from the spending of exhibition organisers, exhibitors and visitors. Money is spent on travel, accommodation and leisure activities.[49] National and international exhibitions are particularly beneficial as many visitors and exhibitors stay overnight at the city for several days. As is the case with conferences, the ancillary spending generates business for a wide range of suppliers who operate from the city. Cities also use exhibitions to promote an image of a centre of industrial and business excellence and expertise.

*Chapter 7*

# Walking and the Tourist Experience

## 7.1 THE NATURE OF WALKING

'Passegiare' is a common activity among Italians.[1] The older part of Italian cities grew and developed around this ancient activity. 'Passegiata' literally translates to 'going for a walk', but it has a much richer meaning and includes the idea of strolling for leisure and enjoyment. Pavia[2] describes his own experience of going for a 'passegiata' with a member of his family as follows:

> We walk for hours. We have no particular destination in mind and we let ourselves get lost in the city. Like two flaneurs, we wander in the labyrinth of the old fabric. As we walk through the narrow streets and plazas, the city lures us even deeper into the core held by its monumental walls, walking far beyond the typical quarter-mile threshold.

O' Mara[3] notes that he has often walked Italian towns and cities and says that he is struck by the wonderful tradition of the 'passegiata' which he describes as 'the sociable, evening stroll along the streets, greeting and chatting with neighbours and friends'. Landry[4] describes it as an urban ritual 'where you look and are looked at, you have an idle chatter and you check out who you fancy'.

For centuries, it was through walking that people perceived and understood space. In the past century, the increased use of the automobile and mass transportation changed all that and introduced new ways and urban forms to perceive and understand space. Today, in large parts of many cities, walking has become a marginal activity to the extent that the design of urban space discourages walking. Since the middle of the past century, cities have been constantly designed and reengineered for motor vehicles. The outcome is often a rational yet generic and soulless built environment lacking spatial identity and a sense of place.

Most tourists will engage in a certain amount of walking during their visit to a city destination. For some, it may be a key element of the visit. Exploring on foot is a common visitor pastime in cities and can enhance appreciation of the place and its inhabitants, therefore contributing to their overall satisfaction. The experience of a historic urban area is reliant on walking, or more specifically the ability to move around the streets and piazzas.

Walking may seem to be a mundane activity but there is much more to it. Walking is not simply a means to traverse urban space but also a way of becoming acquainted with space. People engage their senses when walking in urban places. Physical movement through space, together with sight, enable people to understand places as three-dimensional entities and develop a strong feeling for spatial qualities with 'a sense of direction, geometry, perspective and scale'.[5]

When visiting an area with several visitor sites, the visitor will inevitably need to move from one site to another. If the sites are in close proximity, then walking is the means to do this. This makes walking an essential element of the tourism activity. Specht[6] claims that tourists walk ten kilometres per day through urban spaces, spending two-thirds of the day in open areas of the city. Tourists can best appreciate the city's aesthetics if they are pedestrians, giving them the possibility to walk through spaces in relaxed and irregular rhythms.[7] Walking is an activity that is intrinsic to sightseeing.

The movement of a visitor in the urban environment is one of discovery and appreciation; tourists will wander and linger taking in the surroundings rather than seeking the shortest route. In spatial terms, urban tourism is a predominantly external activity. In historic towns, much of the tourist activity takes place in the external realm. The primary attractions are the narrow winding cobbled streets, captured vistas, glimpses through urban fabric, textures created by architectural facades, open squares, piazzas or river fronts.[8] The urban heritage exemplifies "the human scale, individuality, care and craftsmanship, richness and diversity that are lacking in the modern plastic, machine-made city with its repetitive components and large-scale projects."[9] A hierarchy of street patterns and open spaces combine to create an attraction that visitors can enjoy. An added attraction of many historic areas is the fact that they continue to be lived in. It is the 'life' within that is as much the attraction as the physical qualities of the setting.[10] The visitor arrives and moves through the external space. In larger historic areas, it is possible for a day visitor to spend an entire visit in the external domain.

Wunderlich[11] describes walking as an essential mode of experiencing urban space but distinguishes between 'discursive walking' and 'purposive walking', with the former being spontaneous and characterised by varying pace and rhythm. Walkers 'consciously explore the landscape while sensorially experiencing it'.[12] In discursive walking, people are well aware of the external environment and participate in it. Discursive walking is a participatory mode of walking during which the walker half-consciously visually explores the surroundings while sensorially experiencing them through sight, sound and sometimes even smell. In contrast, in 'purposive walking', getting there is more important than being there, with the walker taking little note of the surroundings. These observations are relevant to our understanding of the tourism experience because 'discursive walking' is inevitably the mode of walking which tourists adopt and that allows them to use their senses to be aware of and explore their surroundings.

The concept of experiencing space through movement is mirrored by Sinha and Sharma[13] who argue that by means of design interventions, a site can be interpreted so that movement is not just for reaching a destination but also for experiencing the landscape through all the senses; engaging the mind and leading to a complete – rather than fragmented – image. Staiff[14] observes that when walking through a heritage landscape, the viewer is a mobile subject who is seeing the place in constant motion. He notes, 'Movement means constant changes of perception and perspective, constant changes to me in relation to the material, . . . changes in mood and tone, shadow and light, textures, colours, smells, air movement and sounds'.

Beyond sightseeing, motion through urban space provides another layer to the experience of the city. The physical sensations associated with traversing a path and avoiding obstacles can widen the sensory experience.[15] The tourist has to walk around and explore to become aware and experience the mundane elements of the city. In research on Valletta[16] interesting observations were made on how tourists engage with stepped streets. Respondents noted that walking up or down these streets is a different experience to walking on level ground. It involves greater exertion and, in a sense, also greater interactivity between the tourist and the steps, as one has to watch where to place one's foot. Stepped streets signify steep slopes and steep slopes signify vistas down the street. In walking up and down the steps, the tourist is doing two things; namely watching their step and looking at the surroundings. The latter entails observing interesting features in the surrounding streetscapes (balconies, architectural detail, doors, windows and so on). It also involves observing the vistas up and down the street and noting how the views change at different levels. Even if the exertion is greater, the increased interactivity of this kind of walking enhances the experience and makes it more enjoyable.

Apart from visiting paid visitor attractions, in a city destination, a tourist will spend time walking and exploring in historic areas in streets, squares and also in city parks. In a tourism context, discursive walking as described by Wunderlich[17] could also be termed as leisure walking. Apart from the spontaneity and the varying walking rhythm, in leisure, walking there is the specific intent of engaging in walking for relaxation or relieve stress. In leisure walking, some of the time might also be spent sitting on public seating in urban spaces; if nothing else then at least to rest their feet. In many cities, locations on high ground from where pleasant views can be enjoyed are often popular with tourists. The views could be across a river, over parts of the city or of open countryside. Taking photos is often part of the leisure walking of the tourist; often stopping to take a photo of anything that they consider visually pleasing. For some, the photo is not just a reminder of the physical aspect of the place but also of the feelings felt during the experience.

Occasionally, tourists stop for coffee or for lunch. Often it is not just about the drink or food but more about the ambience and surroundings. In a Mediterranean climate it is enjoyable to have a drink outside in pleasant weather, especially if it is an area dedicated to pedestrians. Having sea views or views of pleasant scenery makes it even more enjoyable. Consuming local food or drink gives the tourist a better feel of the local character and therefore generally enhances the experience. For outside tables and chairs people often engage in another activity namely that of people watching. When in a city centre, it is not uncommon for tourists to spend some time shopping, even if this would be ancillary to the sightseeing.

The distinction between 'leisure walking' and other activities is sometimes blurred. For example, the tourist may decide to walk from the accommodation to the paid visitor attraction but to do so in a leisurely manner and enjoying the street atmosphere or the urban heritage along the way.

There are many factors that will influence the time a visitor spends in leisure walking; some relating to the visitor and others relating to the destination. It can be safely said, however, that virtually all leisure tourists to a city destination will spend some time walking in the urban spaces. This time is an important component of the visitor's experience of the city. It is reasonable to assume that the time spent in this manner is probably two or three hours per day, possibly more. This is a significant amount of time. In spite of this, the relationship between walkable urban space and tourism has received little attention in tourism literature.

In a similar manner to historic areas, walking is an essential part of the visit to historic monuments or archaeological sites that are sprawled over a larger area. Examples that come to mind are the Acropolis of Athens in Greece (figure 7.1), Pompeii near Naples in Italy and Edinburgh Castle in Scotland. Visitors are engaged with the site being viewed and experienced. They actively seek things that interest them either because they are visually different or because of the site's narrative. The visitor explores a site and seeks enjoyment from the act of exploration and eventual discovery. Moving around, and hence walking, is intrinsic in the visitor's experience. For some,

**Figure 7.1** A visitor to the Acropolis in Athens needs to do a lot of walking to fully appreciate the heritage and aesthetic values of this important site. *Source:* Author's own.

there is also a sense of adventure in discovering things that the tourist has never seen before.

## 7.2 TOURIST MOTIVATIONS AND WALKING

Tourist motivation is related to human needs in that the satisfying of human needs can be a motive for tourist behaviour. Human nature involves a set of universal basic needs with differences between these needs on an individual level leading to the uniqueness of personality.[18] In his work on the psychology of tourism, Ross[19] identified those needs that may be applicable to tourist behaviour. Out of twelve needs identified by Ross, there are three needs that can be considered particularly relevant to walking in a historic context.[20]

(1) 'Cognizance' is described as 'To explore. To ask questions. To satisfy curiosity. To look, listen, inspect. To read and seek knowledge'. The experience of a tourist in an urban environment involves wandering, lingering and taking in the surroundings resulting in exploration and discovery. With exploration comes mystery: the promise for further information. Pursuing 'mystery' leads to discovery which involves the acquisition of new information.

In a historic area, the 'cognizance' motive is the one most likely to be sought and satisfied because the built heritage provides ample scope for the tourist to explore, seek knowledge and satisfy curiosity. The information and meanings offered by urban heritage provide opportunities for the visualisation of the past to aid in the enjoyment of the narrative.

The element of surprise impinges on the tourist experience more than is generally acknowledged in academic literature. Suvantola[21] describes his delight when, after a couple of hours of strenuous walking along a creek, he came across a waterfall that cascaded into a pool. This 'spontaneous encounter' came to him and his travel companions as a total surprise. He points out that the 'most intense aesthetic experience of nature' is likely to catch one by surprise. He compares this experience with another where his group was told that they would be visiting three waterfalls along a road. They went to the waterfalls one after the other by car. The intensity of their experience was nowhere near the intensity of the one they had when they accidentally found the waterfall.

The act of exploration is about seeking out things that are largely unknown to the explorer. Referring to heritage trails, Hayes and Macleod[22] note how visitors are 'invited to "explore" and to "discover" for themselves: personally to find surprises or "hidden" worlds'. When, in the act of exploration, something new or different is discovered, there inevitably is surprise, even if limited. In some instances the surprise will be more significant and this will give rise to a more intense and enjoyable experience.

(2) Linked to cognizance there is 'achievement', described as 'to overcome obstacles. To exercise power. To strive to do something difficult as well and as quickly as possible'. This refers to exploration and the satisfaction derived from discovering something that is different or unfamiliar. The discovery is in itself an achievement. In a built heritage context, the 'discovery' could be an unusual feature or building or an interesting narrative linked to a site. There are numerous elements that are different or unfamiliar to most tourists and therefore a tourist walking in a historic area will have ample opportunity to satisfy any 'achievement' motive they may have.

(3) The third is 'play', described as 'to relax, amuse oneself, seek diversion and entertainment. To have fun, to play games. To laugh, joke and be merry. To avoid serious tension'. The experience of urban space is one of relaxation and feelings of peace and quiet. Tourism is also a means of getting away from the daily routine, including being in surroundings that are not the daily norm and thus allowing the person not to be pressurised or stressed. I would argue that, in a historic area, the 'play' motive is relevant but possibly to a lesser degree as the built heritage provides the context, rather than the object, of 'play'. Some tourists enjoy walking in a historic context without necessarily

referring to the meaning offered by the heritage. It is a context where they can relax and 'avoid serious tension'.

Specific needs are more important to some than to others. Although each need is unique, there are commonalities. A behaviour may meet more than one need. To use a tourism example: walking around and exploring could meet the need of 'cognizance' (in the satisfaction of curiosity) as well as that of 'achievement' (in the satisfaction of discovering something new).

## 7.3 WALKING THE CITY

Many associate tourism activity with the visitation of attractions, but there is more to it than that. Even if not expressed explicitly, many tourists seek to get to know the city they are visiting. O' Mara[23] makes the compelling argument that the best way to get to know a city is by walking:

> Walking a city is the best way to get to know it. You can't get to know the mood of a place, its energy and pace, when you're driving or being driven around. On foot you are directly in touch with city life in all its dirt and glory: the smells, the sights, the thrum of footsteps on pavements, shoulders jostling for position and placement, the street lights, the snatches of conversation.

Before the advent of the car in the early years of the twentieth century, walkability was an essential feature of cities. Everyone depended on ready access by foot or slow-moving carts for access to jobs and the marketplace. The density of people and dwellings was relatively high as everything had to be connected by a continuous pedestrian network. Cities of the middle ages typically packed all necessities of urban living into an area of two or three square kilometres. The same can be said of cities of the nineteenth century as most workers did not have access to horse-drawn carriages or streetcars. As from the 1920s, every development in transport technology brought with it a negative impact on the pedestrian environment.[24] The walkable city was no more due to high-speed transport and the quest for efficiency. Streets were designed to facilitate the fast movements of cars often thereby creating hazardous conditions for pedestrians and imposing barriers to their free movement.

In far too many cities, walkability is something of an afterthought as managing vehicle flow is prioritised. O' Mara[25] argues that city planners seem to conceive people as being contained in boxes: moving boxes (cars) and static boxes (buildings). Walkability is reduced to transition zones between these boxes. People spend most of the time in cars, buses, trains and buildings and relatively little time with the air and natural light on their faces. Some cities are awkward, uncomfortable, exhausting and even dangerous to the walker.

The term 'walkability' is a useful descriptor on how we can walk around in a city. Comparisons of walkability between cities can be made by means of a 'walkability index'. Hall and Ram[26] refer to walkability as 'a multi-dimensional concept that can be broadly defined as the extent to which an environment, usually the built environment, enables walking and is pedestrian friendly'. An alternative and more detailed definition is offered by Southworth.[27] He describes walkability in cities as the 'extent to which the built environment supports and encourages walking by providing for pedestrian comfort and safety, connecting people with varied destinations within a reasonable amount of time and effort, and offering visual interest in journeys throughout the network'. Walking is facilitated by the presence of certain attributes such as connectivity, linkage with other transport modes, safety, quality of hard and soft landscaping, width of pathway and visual interest. A highly walkable environment is one that encourages people to walk by means of a well-connected path network that provides access to everyday places people want to go to. It is safe and comfortable for people of all ages and different levels of mobility. The pedestrian network links seamlessly with modes of public transport such as buses, light rail and trains.

Walking paths should not only support walking for utilitarian purposes, such as shopping or the journey to work, but also provide for pleasure, recreation and health. Walkability is not just about practicalities and physical ease of movement, but incorporates the idea of pleasure.[28] Walking is made more enjoyable if the pathways offer varied visual experiences. A walk that is pleasurable and offers changing scenes and social encounters is more likely to be repeated than one that is boring or unpleasant.[29]

Providing an appropriate level of walkability is a challenge for many city authorities as they try to balance the demand of various users and uses of public space. Urban spaces are shaped by two professional disciplines in particular; urban design and transport planning. Urban design is primarily concerned with the visual and experiential qualities of spaces and normally falls with the remit of architecture. The latter is concerned with people mobility and the ease with which they can move from one place to another, especially within cities. It deals with different modes of transport, including rail, trams, buses, lorries and cars. Walking and cycling are also modes of transport and therefore fall within the remit of transport planning. The layout of roads and pedestrian routes within cities are designed by transport planners. Over the past century, in many countries across the world, these two disciplines have evolved along different tracks with urban design focusing on the experiential qualities of the built environment, whereas transport planning focusing on functionality and efficiency, particularly for the motorist.[30] This was less so in some European countries. As from the 1980s and 1990s, many European cities invested in better public transport system (particularly buses and light rail). This facilitated the creation of urban spaces that are more amenable to pedestrians.

As a mode of transport, walking has limitations of distance and speed as well as exposure to inclement weather. It has, however, numerous advantages. Cities should be made more walkable as this will derive many benefits to residents and to visitors. Encouraging walking and cycling reduces the pressure on the infrastructure of other transport modes. Walking and cycling are environment-friendly as they do not involve the consumption of carbon fuels. They do not pollute the air and noise emissions are minimal. Most cities have adopted policies that stress the importance of public transport and pedestrian and cyclist mobility, partly in response to the problems of overcrowded road networks which are struggling to cope with rising car ownership.

More than a purely utilitarian mode of travel to work, school or the store, walking can have both a social and recreational value. It is socially equitable as it is available to the majority of the population, irrespective of age and financial means. The children, the elderly and the poor suffer disproportionately from living in cities that are oriented towards car travel. The reason is that they are most dependent on other forms of transport.[31] The residents of walkable cities are more likely to have better physical and mental health. Among the health benefits of walking are improved cardio-vascular circulation, reduced stress, better weight control and improved mental alertness. Walking is the most accessible and affordable way to get exercise and having a walkable city is the best way to encourage residents to walk and engage in other forms of light exercise. Whatever the travel, everyone is a pedestrian at some point because walking is part of every trip. The need to enhance walkability is even greater in central areas of cities and areas that are close to public transport stops and car parks.

The concept of walkability and catering for pedestrian needs is frequently linked to broader sustainability goals embracing ideas of liveability, active mobility, accessibility and efficiency of public transport and reduction of private car use. A city authorities' effort to make streets more walkable is normally intended for locals as part of an overall strategy to make the city more liveable. Even if not directly intended for tourism, efforts to improve walkability will also greatly benefit tourism. Some cities have been very successful in this as their streets and urban spaces 'have a porous and fluid quality that makes walking around them a joy'.[32]

Earlier on in this chapter we discussed how walking is an essential element of the tourism activity as visitors to a city destination are likely to spend significant time walking from one interesting site to the next or simply exploring. It therefore comes as no surprise that the most successful tourism destinations are those that offer ample opportunities for walking. Walkable places are often considered as attractive for locals, visitors and tourists alike. Two examples that come to mind are the South Bank and Bankside in London (Case Study WK1) and Ramba de Mar in Barcelona. Walkability makes a city destination attractive and concurrently extensive areas dedicated to pedestrians enables the city destination to better cope with the numbers of tourists that will visit. In this context, it is surprising that relatively little has been written on walkability's contribution to tourism.[33]

## Case Study WK1: Bankside and the Millennium Bridge, London

The Bankside on the River Thames is an attractive length of riverside promenade stretching about two kilometres from Westminster Bridge to the Millennium Bridge and beyond. The Thames has been a recreational resource for centuries but, up until the middle of the twentieth century, industrialisation and the water transport of goods took precedence over leisure activities.[34] The regeneration of the Bankside took place over decades with major developments and attractions being created at different points in time. The Royal Festival Hall was inaugurated in 1951 and then renovated in 2007 as part of the overall regeneration of the area. In 2000, there were two important additions; the London Eye and the Millennium Bridge.[35] In terms of walking the latter is a vital addition as it connects the Bankside, with all its various attractions, to the major tourist attraction of St. Paul's Cathedral. The bridge has an innovative steel suspension design, in part to enhance the experience of walking across the river. The bridge's alignment is such that pedestrians on the bridge will get a view in the distance of St Paul's Cathedral south façade framed by the bridge supports. On the opposite side of the river, pedestrians can enjoy views of Tate Modern, Bankside and the various buildings and features alongside the riverside promenade. The bridge opened in 2000 with 160,000 people visiting over the first weekend. It was however closed soon after as pedestrians were experiencing a swaying motion that was causing discomfort and unease.[36] It reopened in 2002 after modifications to stop the swaying. Its closure made it very controversial with accusations of waste of public funds. As it turned out, however, the intense public debate made the bridge more widely known something that contributed to its popularity in subsequent years (figure 7.2).

There are also several interesting features and activities that are of interest to the visitor including views across the river and from Millennium Bridge, sand sculptures at the water's edge, buskers, cafés, restaurants and a small shopping arcade.

**Figure 7.2 The Millennium Bridge is a popular pedestrian bridge across the River Thames, London.** It connects the Bankside on one side to the pedestrian road leading to St. Paul's Cathedral on the other. *Source:* Author's own.

The Bankside is a tourist space that illustrates the attractiveness of walking as an activity for visitors and locals. The riverside walk is part of a longer walk that connects two major tourist foci the Big Ben/Houses of Parliament to the west and St. Paul's Cathedral to the east; the latter being reached after crossing the pedestrian Millennium Bridge. Along South Bank and Bankside, there are several places of interest and attractions (London Aquarium, London Eye, Jubilee Park, the Royal Festival Hall, the National Theatre and Tate Modern). In spite of these various high profile attractions research[37] shows that the most popular activity amongst visitors was going for a stroll, followed by visiting a restaurant, café or pub. According to Maitland and Newman,[38] walking is key to the way visitors explore and perceive the area. The presence of a large number of local people and workers making use of the area makes Bankside seem less touristy than other major tourist spaces in the city.

The success of the area is attributable to a number of factors. Walking aside, there are many reasons why people would want to go to the area. There are residences, offices, major theatres, visitor attractions and catering establishments. More crucially, however, the underground makes any part of central London easily reachable by large numbers of people living or staying in London. Using public transport, South Bank and Bankside can be reached by millions of people easily, cheaply and in a relatively short period of time. A third factor is the attractiveness of the area gives good reason to people to come for a stroll along the River Thames.

Cities are large and complex entities comprising a mix of districts of greater or lesser appeal for tourists. Apart from the availability of visitor attractions, the extent of walkability of a city area will influence its overall appeal to tourists. According to one study,[39] the key predictors of tourist walking decisions are safety, comfort, environmental quality and potential for activity and exploration. On any particular day of a tourist's visit at a destination, where to walk and the extent of walking is determined by a range of variables such as weather, time of day, duration of stay, age, physical fitness, interests, gender, travelling companions and personality.

There is a distinction to be made between walking on an organised itinerary or tour (with a guide or using a mobile) and wandering the streets at random. The latter involves greater risks of getting lost or not finding anything interesting but it can also provide greater satisfaction in terms of the surprise of 'discovering' something different that the wandering tourist would have never seen before.

Cities and towns across Europe are home to a wide range of walking trails that are focused on heritage. They are a vital resource for urban recreation for both locals and tourists. There are several types of urban heritage trails but the most common are self-directed interpretive walking tours which thematically link places of historical and cultural interest.[40] Brochures, maps, podcasts and on-street signage enable the tourist to follow the trail though the historic area by providing information, interpretation and promotion. More often than not urban trails are associated with a particular theme. The theme could be a particular period of the town's history or places that are associated with a famous resident's life. Other examples of themes are an architectural style, an ethnic neighbourhood, nationalistic heritage and public art.

An appropriate theme for a walled town is the defensive fortifications and the historic narratives associated with them, including possibly past attempts to invade the city. Town walls may be capable of being walked upon. A walk along the town fortifications provides a flexible product that can be enjoyed at the pace and duration determined by the visitor with the possibility of joining or leaving the fortification walk at any point.[41] The elevated walkway often offers spectacular outward views of the surrounding countryside and seascape, as well as inward views of the town itself. Fortified towns are generally small and walking along the surrounding fortifications allows for the 'gem city to be observed and experienced holistically'.[42] Depending on width and other factors, an extensive circular walkway on the town walls can be created which Ashworth and Bruce[43] describe as 'an almost ideal tourist experience'. Chapter 4 discusses fort and fortifications as visitor attractions.

Trails are also provided at heritage sites and visitor attractions that are sprawled over a large area. Examples include archaeological sites, historical parks, outdoor museums and heritage gardens. Most outdoor heritage sites have some sort of planned route that leads visitors from one location to the next. Trails are useful management tools as they facilitate interpretation, guide visitors away from the more sensitive parts of the site, facilitate access for people with mobility difficulties and provide vistas and photo opportunities at various points of the property.[44]

The experience of a historic area is not limited to walking and exploring. It is normal for a tourist to visit at least one museum or a cultural attraction, normally against payment. The number of such visits is dependent on a range of factors, including the background and aptitudes of the visitor and the cultural orientation of the city being visited. Many cities have rivers running though their central areas. Boat trips offer tourists the opportunity to view the main sights from a different perspective with the added advantage of sitting comfortably and being provided with the information. The closeness to water enhances the experience. After a lot of walking, tourists need to rest their feet and therefore they will also spend time sitting normally in a quiet place in a garden or in a busier place within a pedestrian area.

## 7.4 PEDESTRIAN AREAS IN CITY CENTRES

Pedestrian streets are a common feature in the commercial and historic centres of cities across Europe. The configuration and extent of pedestrian areas in a city centre are dependent on many factors including street layout, street widths, access by public and private transport, land uses and city governance. In its simplest form, a pedestrian area consists of a single street with several side streets. Alternatively it could be two or more streets that interconnect with each other, often through a square, to form a continuous and longer lineal street. In almost all cases, the ground floors of buildings provide a continuous frontage of lifestyle and fashion shops, department stores, cafés, restaurants and banks. City centre pedestrian areas often include important civic and religious buildings such as city halls, churches and museums – some of which would also be of significance in the city's history. It is not uncommon for part of the pedestrian area to be dedicated to an open market that operates weekly or sometimes daily.

Pedestrianised streets are key to improve the environmental quality for pedestrians and making communities more liveable, particularly when they are well designed, managed and strategically connected to networks of public transport, pedestrian paths and bike routes. Pedestrian schemes must be carefully planned to be successful and achieve the desired quality. They must be attractive to pedestrians; allowing for the discovery and appreciation of the environment as well as be inviting. In pedestrian streets, appropriate street furniture is used like, for example, seating, trees, artworks, fountains, benches, paving and other features – features which make the town centre's streets a more sociable space.

During the 1950s, pedestrianisation in Europe was small-scale and was at most limited to a single street. There were many urban planners who were calling for more streets to be pedestrianised, but they were opposed by hard-core traffic engineers who were concerned that areas without cars would enter into a downward cycle of lack of investment and deterioration.[45] By the late 1960s, there were rising concerns that the urban environment and urban life were increasingly being dominated by the car. It seemed that modern life meant that cars had to be allowed everywhere and that cities had to provide for them with more and more space being dedicated to moving and parked cars.

Attitudes to pedestrian streets and to motorisation changed when a new style of pedestrianisation was demonstrated in Munich in 1972. Extensive pedestrianisation of the city centre was one of the projects linked to the 1972 Olympics. The move was resisted by some transport experts who argued that the area was so large that pedestrians would feel isolated in unfriendly streets and thus stay away. The opposite happened. Munich's newly pedestrianised streets turned out to be hugely successful with an impressive increase in the number of people visiting the city centre to shop and for other purposes. Munich's success encouraged other cities in Germany and across Europe to implement larger pedestrianised areas in their commercial centres.[46]

An erroneous view is that the creation of pedestrian streets simply involves the exclusion of cars. The right approach to pedestrianisation is that it should be part of an overall transport strategy for the area which considers other modes of transport; particularly vehicle traffic and public transport provision in and around the area to be pedestrianised. In many cities, the extension of the network of pedestrian streets is accompanied by significant investments in the public transport infrastructure, most notably light rail. Public transport provides access to the heart of the city.[47] Other measures could include traffic calming in secondary roads, establishment of cycle routes, the availability and provision of off-street parking and the introduction of park and ride services. Normally cycling is allowed in pedestrian areas, except where pedestrian numbers are high as the mixing of crowds and cyclists may result in a pedestrian being hit by a bicycle. A downside of extensive pedestrianisation in city centres is the reduced access to residences. Another disadvantage is the increased demand for car parking at the immediate periphery of the pedestrianised zone, thus making car parking more problematic for residents. In commercial streets, arrangements are made for the servicing of shops with delivery and service vehicles being allowed in at specified times, normally before 10 a.m.

The larger the pedestrian area, the greater the need for the pedestrianisation scheme to be part of a comprehensive transport strategy. A large-scale pedestrian area is not possible unless there is an effective public transport system that allows easy access to the pedestrian area to as many people as possible. Large-scale pedestrianisation with poor public transport access will decrease, rather than increase, activity within the town centre. When an urban area is being pedestrianised, public transport will need to be reorganised to have stops as close as possible to pedestrian streets. There may be situations where public transport vehicles are allowed to pass through the pedestrian areas. With light rail, this can be an effective way of retaining good access to the city centre and thus enhancing its commercial viability.

Most European cities of a population between five hundred thousand and 1 million have public transport systems based on tram or light rail, backed up with a bus network.[48] Many of the tram systems were first established in the latter years of the nineteenth century or the early years of the twentieth century. Light rail is a more recent transport technology that has been introduced by some European cities in the past few decades. Larger cities have developed underground metro systems: the larger the city, the more extensive the metro.

Apart from public transport, the city centre has to be made easily accessible to users of private cars. This is best achieved by locating multi-storey car parks strategically with easy access by car from the main road network and, concurrently, easy access on foot to the pedestrian streets. Apart from being expensive, finding suitable sites for multi-storey car parks is difficult in most cities so an alternative approach would be the provision of a park and ride service with the surface car parks being at the periphery of the city.

For environmental reason, it is good practice for city authorities to encourage people to use public transport rather than their private cars to access the city centre. This can be achieved through an appropriate pricing mechanism for public transport and parking. It is not uncommon for citycentre car parking to be prohibitively expensive.

Whenever a new pedestrianisation scheme is proposed more often than not retailers are opposed because of concerns of loss of trade. Experience shows that retail activity will benefit from pedestrianisation.[49] Generally pedestrianisation increases the number of pedestrians. More pedestrians, however, does not necessarily mean higher turnover for all shops. Shops that are less attractive, smaller or located at the periphery of the pedestrian area may lose out.

The closure of some streets to traffic in the city centre will bring about changes to the traffic circulation that will inevitably impact traffic roads in the surrounding area. Traffic that previously used the city centre road is shifted to other streets. Careful planning is therefore required for a road to be permanently closed to traffic including detailed traffic studies. Many cities across Europe were fortunate to have embarked, in the 1960s and later years, on a programme of building ring roads to ease the increasing traffic congestion in city centres. This eventually made it easier to pedestrianise city centre streets as the main flow of traffic was already passing through arterial roads away from the city centre.[50] For cities without good ring roads to city centres pedestrianisation was more difficult as this required measures to decrease traffic volumes in and around the city centre, usually by means of financial disincentives coupled with major investments in public transport infrastructure.

Much of the previous discussion refers to pedestrian areas in city centres. In most city centres, pedestrian streets are the main shopping streets of the city. Such streets are usually crowded with shoppers, window shoppers, tourists and other people of all ages and from all walks of life. Some may be just passing through while others are there to shop, to stroll, to observe, to eat or to recreate themselves. Sitting areas, public art, street performers, outdoor cafés make pedestrian shopping streets lively, attractive and enjoyable.[51] The absence of cars and other vehicles makes them safe. During special celebrations and events such as the Christmas season and weekly market days, pedestrian shopping streets are particularly busy and vibrant.

An interesting intervention in a shopping street in Rotterdam's city centre illustrates the importance of creating continuous shopping spaces within the cityscape. Two concentrations of retail areas were separated by a busy traffic road; creating the inconvenience for shoppers to wait at the traffic lights to cross from one retail area to the other. The solution to this was the excavation of an underpass below the traffic road. The underpass and the gentle ramps leading to it are lined with shops and are well lit.[52] Walkers are barely aware that they are at one level below street level. This was a new retail complex nicknamed the 'Koopgoot', or 'shopping trench'.[53] Completed in 1993, this intervention joined two popular retail areas into one. It created a lively shopping area dominated by people and without cars.

Although pedestrian streets are normally associated with a retail function, there are many different types of pedestrian areas depending on the uses (residential, shopping/ commercial) and on the historic/cultural value of the area. There are also streets where the pedestrian is given priority even if limited access for cars is allowed. These are referred to as pedestrian priority streets.

Street furniture and lighting is an integral part of urban space design and of pedestrianised streets. In historic areas, rather than try to recreate an 'authentic' that never existed, the design should seek to cater for contemporary urban life while respecting the historic context. The choice and use of surface material are also important. In new urban developments, concrete paving blocks are often used. In a historic context, the use of cobblestone is more appropriate and in keeping with the context. Cobbled streets may, however, be problematic for people with mobility difficulties so innovative solutions are required to cater for different levels of mobility.

Pedestrian streets are relevant to tourism for a number of reasons. They provide the context within which the tourism activity takes place. For example, in historic areas, pedestrian streets enable visitors to view the various historic buildings and features without the nuisance and reduced safety of passing traffic. If the main visitor attractions are in the city centre, it is more pleasant to walk from one attraction to another along pedestrian streets. A pedestrianised city centre can be an attraction in its own right as visitors enjoy the activity and the buzz that is created by the urban activity within them. Many city centres have rivers or water canals running through them or are facing directly onto harbours. The pedestrianisation of the waterfront provides an attractive urban environment because of the outward views over water – an environment that is made lively and sociable with the presence of cafés, bars and restaurants (figure 7.3).

Figure 7.3 A popular pedestrian urban space alongside a historic canal at Nyhavn, Copenhagen. *Source:* Author's own.

Whether moving traffic or parked cars, the car is often seen to be intrusive in a historic context. Pedestrianisation is often associated with urban renewal and conservation of historic areas as it creates an atmosphere that is more amenable to the historic context. Moreover, it allows people to appreciate the historic buildings surrounding the space. This was the case in Merchants Street in Valletta, Malta. The street includes some of the finest historic architecture in Valletta but this could not be appreciated by Maltese and tourists because almost the entire width of the street was taken up by parked cars and moving traffic. Pedestrians were relegated to narrow footpaths on either side and even those were sometimes partially blocked by parked cars. All that changed when the street was pedestrianised in 2007. Another vital advantage of pedestrianised spaces in historic areas is the removal of a major source of pollution and hence historic buildings are better protected. A mix of rainwater and pollution significantly increases the rate of deterioration of stonework of historic facades.

For decades, the trend across Europe has been to create pedestrian areas in town and city centres. For centres that were already pedestrianised, the trend was to extend them further and create urban spaces with more and more spaces dedicated to pedestrians.

There are different ways how pedestrian-oriented urban spaces can come into being. The first and most obvious are those spaces to which access by vehicles is physically not possible because they are too narrow or because they are steeply sloped or stepped. The second is when physical obstructions are made to prevent cars from entering the space. This would necessitate careful traffic planning to ensure that the diverted traffic will not create problems elsewhere. The provision of multi-storey car parks is recommended to compensate for the on-street parking that would be lost. More crucially, there needs to be efficient public transport so that this will be a viable alternative to the use of the car. A third approach is to create a public space out of reclaimed land from the sea. In protected waters, this could take the form of timber decking. On more exposed coastlines, a harder more durable landmass would have to be created. A fourth approach would be the conversion of abandoned or derelict land into a public space or garden. This could be part of a wider regeneration programme of a disadvantaged area of the city. A fifth approach is for the grounds of an existing structure or facility to be opened to the public without charge. This could be under-utilised land near a church, a monastery, a stately home or a fortified landscape. Allowing free public access would come at a cost for security and maintenance. On the other hand, such spaces are more likely to be of interest to tourists and would therefore generally enhance the tourist attractiveness of an area. It could also be a means for better tourism management for the relief of pressures on nearby tourist hotspots.

Case Study WK2 illustrates some of these different approaches to creating spaces for pedestrians.

### Case Study WK2: Walkable urban spaces in Barcelona

The historic districts of El Raval, Barri Gòtic, Ciutat Vella and La Ribera in Barcelona provide an extensive network of streets and piazzas that are predominantly for the use of pedestrians. This network did not come about by chance but with consistent urban policies in favour of public transport and the pedestrian over the private vehicle. Public transport is efficient and reliable across the city including in the historic cores. On-street parking is minimal and only against payment. A few streets in these areas carry traffic but the volume is low.

From a tourism perspective, this makes these districts more interesting and enjoyable to walk and explore. In Barcelona, it is not just the tourist hotspots that are enjoyable (such as La Rambla and Plaça Reial) but also those in the more residential areas (such as La Rambla del Raval and Plaça de Vicenc Martorell). Some streets are pedestrian only because their width does not permit traffic. Others are wider and could allow traffic and parking but instead have been dedicated to pedestrians. In some cases, facilities for residents are provided such as children's playgrounds and enclosed spaces for dogs.

Pedestrian oriented spaces were also created in other parts of the city. There are the monumental Plaça de Carles Buïgas and Plaça de les Cascades leading up to Museu Nacional d'Art de Catalunya. The area is extensive with several interesting features and adjoining buildings. This allowed the designers to create urban spaces that are amenable for leisure and relaxation. The same can be said for nearby Plaça d' Europa in the Olympic sports complex. Another example of a pedestrian only urban space is the road, Av. Gaudi, connecting two important cultural and tourism sites; la Sagrada Família and the Hospital de la Santa Creu i Sant Pau. Barcelona also has its own sandy beach at La Barceloneta and alongside it is a promenade stretching several kilometres long.

A clever intervention was the creation of Rambla de Mar; a pedestrian walkway across the entrance to a yacht marina connecting the city centre waterfront (at the end of La Rambla) to a major shopping complex at Plaça de l'Odissea. The walkway is approximately 200 metres long and is made of timber decking. What used to be sea space for boats was converted into a space for people. At pre-determined times, a short stretch of decking swings open to allow yachts to enter or exit the yacht marina within. Attractive design and the connection of two major foci of activity makes this space busy with pedestrians. Plaça de l'Ictineo is another interesting space along the city centre waterfront. This is a small park whose layout and attractive design maximizes its leisure potential.

126                           Chapter 7

This extensive network of pedestrian-oriented spaces came about because of a consistent commitment by the city authorities over several decades. These various initiatives cannot be seen in isolation but should be considered in a wider context of urban and cultural regeneration. The Olympic Games of 1992 in particular were an opportunity for the city to invest in public spaces and also to rebrand itself.

Barcelona offers a diversity of walkable urban spaces. Some are small and intimate. Others are monumental. Some are in a historic context while others are surrounded by modern development. Some are in an urbanised context. Others look out over the harbour or open sea. Still others are characterised by greenery. The diversity of walkable urban spaces is part of the attraction of the city. Barcelona's pedestrian-oriented urban policy has provided many benefits to pedestrians and hence also to the tourists. The likely objective of the authorities was to create a city that is more liveable for its people but, in so doing, they have also created a city that is more enjoyable to walk and more enjoyable for the tourist. Barcelona's success as a tourism destination goes beyond its many outstanding visitor attractions. Its success is intricately linked to the extensive availability of attractive walkable urban spaces (figure 7.4).

**Figure 7.4 Rambla de Mar, a pedestrian walkway on timber decking at Barcelona.** It connects the city centre waterfront to a major shopping complex at Plaça de l'Odissea.
*Source:* Author's own.

## 7.5 PHOTOGRAPHY AND THE TOURIST EXPERIENCE

A practice closely associated with tourism activity is the taking of photos. In urban areas that are of historic or architectural interest, it is not uncommon for tourists to stop and take photos of the buildings or scenes that are in front of them. Documenting justifies and even glorifies the trip.[54] For some tourists, taking photos constitutes a major focus of activity to the extent that it becomes an integral and vital part of their tourist experience.[55]

Up to the end of the 1990s, the practice of photography in tourism was very different from what it is today. Film cameras necessitated the printing of photos for them to be viewed. This was done at a cost so the number of photos taken during a holiday would be relatively few. Since then, there have been three distinct, yet closely interrelated, technological developments that transformed photo-taking and sharing. First was the introduction of digital cameras. These are cheaper, easier to use and more portable, therefore making them more accessible to a wider market.[56] It is relatively easy for anyone with no technical skills to take photos and place them on social networks and other sites. With digital cameras, the tourist is not concerned that they will incur a cost with every photo taken because the cost of 'wasted' shots is nil. This allows the tourist to experiment in taking photos and also to take shots of objects and views which may be considered as unremarkable and commonplace. The subject matter of tourist photography has been extended. Although traditional tourist photos persist, we are now witnessing photographic practices where 'ordinary scenes or banal moments' are turned into 'something noticeable, and thus something recordable'.[57]

The second development was the use of the internet to share digital photos. With the increased popularity of social media, it is now common practice for people to share their photos with friends and family; including photos of their visits to places away from home. The third change was the integration of cameras into mobile phones. Whereas at one time the camera was the iconic accessory of the tourist, today many carry cameras all the time in the form of the smartphone. This also made it easier than previously for people to share photos on social media or through other means.

Digital photography brought about a change in the way that people use photos. There has been a shift from using pictures for purposes of memory and commemoration towards using photos as a form of identity formation and to affirm individual identities and personal bonds. Van Dijck[58] notes how taking photos has shifted 'from family to individual use, from memory tools to communication devices, and from sharing (memory) objects to sharing experiences'. Youngsters share experiences as opposed to sharing memories. The changing nature of photography and its use on the internet and other personal communication devices has changed not only people's photographic practices but also ways of experiencing place.[59]

The act of photographing makes the observer more keenly aware of the physical surroundings. While preparing the photographer will glance at and observe the various items which make up the overall scene. The photographer may stroll around to establish the best location from where to capture the scene and the various elements which compose it. Lee[60] describes it as 'the practice of being attentive to a series of temporal and spatial moments in the locale and trying to make sense of and interpret a subject'. In addition, these practices 'can induce a curiosity for the memories of the locale and the people who have lived there'. In the context of a visitor to a heritage area these photographic practices become all the more relevant. The act of 'finding' interesting features and views to photograph is one of 'discovery' which, for many visitors, is an important and fun part of the tourist experience.

Photography is a medium for framing the extraordinary experience; structuring the experience as well as structuring the narratives of experience. Andersson Cederholm[61] provides an interesting framework for better comprehension of the practice of photography. This is based on three analytic themes. The first, framing the unique, refers to the taking of a picture of those motifs which are regarded as typical for a place in accordance to norms and conventions as to what one should photo when on holiday. Everybody knows what the Eiffel Tower looks like, but when in Paris, one feels almost compelled to take a photo of it. Similarly many visitors to Copenhagen make it a point to go to Langelinie promenade to see the Little Mermaid, a bronze sculpture displayed on a rock by the waterside. The statue is highly symbolic of Copenhagen. Taking and sharing a photo next to the mermaid is the tourists' way of telling friends and relatives that they have visited Copenhagen (figure 7.5). At peak times, the shoreline next to the mermaid is so crowded that it is difficult for anyone to take a photo 'alone' with the mermaid.

The second, framing the local scene, refers to the search for authenticity and the attempt to 'the search for intimacy with the locals'. A tension between 'intimacy' and the touristic consumption of places develops. The third theme is catching the moment. In taking the photo, the tourist seeks to 'freeze and frame the experience'.[62] There is an inherent contradiction in this kind of consumption. With the camera as a tool, the tourist consumes the experience while at the same time seeking to preserve it. The freezing of the moment through the photo is an act that is directed to the future.

The transformation of photo-taking and photo-sharing is also relevant to tourism in that communicating experiences with the help of photographs is increasingly becoming an integral part of tourist photography.[63] Photography is in a sense a 'form of symbolic capture'[64] whereby the tourist captures the

Figure 7.5 Being photographed alongside the Little Mermaid in Copenhagen. *Source:* Author's own.

image of a place with the intention of viewing, appreciating and sharing it at a later time. Similarly, Andersson Cederholm[65] notes that a traveller is 'a collector of experiences, which means that the homecoming rituals of photo shows and story telling, is an important part of the travelling experiences'.

Traditional snapshots are taken at well-known tourist sites and photogenic landmarks for the commemoration of their 'having-been-there'. Tourists often seek out particular views that are considered 'photogenic' or 'iconic' and reproduce them in their photographs. Sometimes tourist discourse gives an impression that certain attractions simply must be seen. Taking photos of the attractions is proof that this has been done. Photography is a form of documentation that a pursued purpose has been achieved. The purpose may be to be involved in a different culture, to experience an uplifting attraction, to view an inspirational landscape or to be socially engaged with a travelling group.[66]

Lee[67] argues that people are now more inclined to capture moments and places which previously used to be disregarded 'such as those moments that are banal but personally intriguing, those events that people experience accidentally or instantaneously on a site, or those routine places where people don't usually carry cameras'. Photography is a means for recording sights that may seem mundane but are in some way is meaningful to the viewer.

Haldrup and Larsen[68] suggest a motive for tourist photography based more on social activities than the desire to consume places. According to this view, tourism is merely the stage for framing personal stories revolving around social relations, particularly among the photographer's accompanying family, which can later be told and re-told through the medium of the photo album or slideshow. Andersson Cederholm[69] notes that, just as it is for birthdays, graduations and other special occasions, photography is 'a tool for framing the extraordinary event of travelling'. For trips involving above-normal expenses, the capture and sharing of experiences by means of photography is also a means of justifying the expense of the trip.[70]

People's perception of places, as well as the photo-taking practices at the tourist sites, is shaped by the image projected by mass media and tour marketers. In turn, the photos taken at the tourist sites perpetuate the iconic images of the places.[71]

*Chapter 8*

# Meaning of Place and the Tourist Experience

## 8.1 THE EXPERIENCE OF PLACE

Visiting a place can evoke feelings and emotions in a person that cannot be explained by the physical properties or sensory perception. This is often referred to as 'sense of place' or 'genius loci'. Montgomery[1] notes, 'It is a relatively easy task to think of a successful place . . . but it is much more difficult to know why a place is successful'. A place is not simply a geographical site with definitive physical characteristics. There is something that goes beyond the physical. People feel better in spaces that have certain qualities.[2]

Different authors provide broadly similar understandings of *genius loci*. Berleant[3] argues that 'place' is neither a physical location nor a state of mind. He describes it as 'the engagement of the conscious body with the conditions of a specific location'. Norberg-Schulz[4] describes 'genius loci' as representing the sense people have of a place, understood as the sum of all physical as well as symbolic values. People use their value systems to integrate built form, topography and natural conditions to form a sense of place.[5]

The perception of space develops from a range of environmental stimuli. Senses play a vital part in our experience and understanding of the environment.[6] A discussion on the tourist experience of urban spaces is incomplete unless the role of senses is adequately explained. Vision is the sense that provides most information about shapes, colour, detail and texture of buildings and features. The senses of hearing and smell also provide environmental information but to a much lesser degree. Senses each have their own range. While many people may be able to see for many kilometres, the senses of hearing, smell, touch and taste are much more spatially restricted. Senses constantly reinforce each other to extend and broaden sensory perception allowing for a broader appreciation of the surroundings.[7,8] Light, particularly daylight, is an important factor that impinges on how people perceive and experience space. There is a link between a good luminous environment and physical, intellectual and emotional well-being.[9] Vision is the most effective because it is active and searching; 'we look; smells and sounds come to us'.[10] The senses are unable to take in all the information they are bombarded with, resulting in the brain selectively filtering out a large amount of environmental information.[11]

Physical movement through space, together with sight, enable people to perceive the three-dimension and develop a strong feeling for spatial qualities. As a person moves through space, they use their senses to take in environmental information and then process the information into images or impressions of the environment through which they are moving. The most widely studied and well-known attempt at understanding and explaining this process is through the method of mental (or cognitive) mapping.[12] This was a groundbreaking technique pioneered and developed by the urban planner Kevin Lynch in his 1960s study 'The Image of the City'.[13] The technique was to collect sketch maps that urban residents produced of their city. Lynch argued that these provided visual representations of the knowledge that residents possessed of their surroundings and were relatively stable, learnt images of the environment that are used to orientate and navigate. The mental maps revealed that people's environmental knowledge was partial, simplified and often highly distorted. People mentally organise the environment into a coherent image, one that enables them to move about easily and quickly through urban spaces. The ease with which this can be done is dependent upon the legibility of the urban space. Lynch[14] refers to environmental legibility and describes it as the 'ease with which its parts can be recognized and can be organized into a coherent pattern'.

From various studies in the Lynch tradition, it seems that some cities are more legible than others.[15] For example, Amsterdam is more legible to its residents than are Rotterdam and The Hague to theirs, probably mostly due to the radial network of roads and canals. Milano and Rome were both found to be very legible but in different ways. In a study comparing the two cities, the mental maps of the Milanese were structured by a clearly connected set of paths related to the city's radial street pattern. The mental maps of Rome's residents exhibited a greater diversity of content and tended to be structured around landmarks and edges associated with the city's historic buildings and the River Tiber.[16]

Lynch found that, while the mental maps produced by individuals were all unique, he could recognise a series of broad similarities between them. The mental maps indicate that there are five key physical elements that support legibility of urban spaces namely paths, nodes, landmarks, districts and edges.[17]

- Movement of people takes place along paths, such as streets and roads.
- Nodes are where paths converge. Busy public spaces and spaces outside major transport interchanges are examples of nodes.
- Landmarks can be seen from a distance, normally from different angles and over the top of other smaller structures. Their form and size make them stand out in unremarkable surroundings. Landmarks are usually unique and easily recognisable.
- Districts (e.g. neighbourhoods) are urban areas within the city-bound by a common theme by reason of building type, uses or some other characteristic. Districts may have hard precise boundaries or soft uncertain ones, gradually merging the district with the surrounding areas.
- Edges are linear elements that act as barriers defining an area and distinguishing it from adjoining areas. Waterfronts are examples of edges that are found in many cities. Whether in a port city or along a river, the shoreline along a waterfront is the physical edge that separates land from water. Another example is lines of fortification walls. Many historic towns have retained at least part of their fortifications. Except at gateways, the physicality of the fortification wall and the difference in levels act as barriers for movement across the edge.

Using Lynch's work[18] as their basis, Sinha and Sharma[19] propose measures how the environmental image of an area can be made stronger including clarifying pathways, enhancing node prominence, developing districts, strengthening edges and preserving the singularity of landmarks.

Montgomery[20] argues that an amalgam of physical elements combines with 'the psychology of place' to produce urban quality. Physical elements refers to the architectural form of buildings and structures as well as features such as scale, vistas, open space, greenery and street furniture. 'The psychology of place' refers to the social, psychological and cultural dimensions of a place. Similarly Carmona et al.[21] note that the significance of the physicality of places is often overstated and argue that activities and meaning may be as, or more important, in creating a sense of place. Whereas physical setting and activities can probably be easily appreciated, meaning is more difficult to grasp.[22]

Perceiving space as 'place' is not just about environmental stimuli. There is also the subjective aspect and the immaterial and imponderable.[23] Place is not a constant in all circumstances. It has 'a range of subtleties and significances as great as the range of human experiences and intentions'.[24] Creswell[25] describes place as a way of seeing, knowing and understanding the world involving a rich and complicated interplay of people and the environment. For Stokowski,[26] sense of place refers to an individual's ability to develop feelings of attachment to particular settings based on a combination of use, attentiveness and emotion. She notes that the same setting can mean different things to different individuals and claims that 'places are also fluid, changeable, dynamic contexts of social interaction and memory'.[27] Where there is no 'emotional union' between the physical form and the visitor, 'the precinct remains a space rather than a place'.[28]

People feel better in spaces which are felt to contain certain qualities. Place-making encompasses both material production and symbolic order and hence an important aspect of placemaking is how places are imbued with meaning. To become meaningful, places must be verbalised and narrated, and various means of communication must contribute stories and images that advance the construction of place.[29]

Urban design is a professional discipline that falls under the overall umbrella of architecture. It deals with the design of spaces in the built environment, including squares, piazzas and streets. The main focus is to create urban spaces that people feel comfortable in and that serve their urban functions. More than that, urban design should go beyond the physicality of an urban space and truly seek to create 'place'.

In urban design literature, sense of place is explained by means of a model with three basic elements namely physical setting, activities and meanings.[30,31,32,33,34] These three components are deemed to be interrelated and inseparably interwoven in experience.[35] Ebejer et al.[36] argue that the intrinsic qualities of the space are relevant to the tourist experience but what is even more relevant are the interactions of the tourist with different elements within that space, namely interactions with self, interactions with others and interactions with surroundings. Loeffler[37] refers to significant and profound effects that the outdoor experience has on individuals and categorises them under three headings. The first is 'inner exploration' as a form of 'self-discovery'. The second refers to 'connections with others' or 'attachment to friends/group'. The third relates to the close interaction or 'spiritual connection' with the environment and the attachment to place. These can be summed up as interaction with self, with others and with the environment.[38]

Referring to form, activity and meaning, Carmona et al.[39] point out that sense of place does not necessarily reside in these elements but in the human interactions with them. Griffin and Hayllar[40] also discuss space qualities in the context of interactions: 'The atmosphere is created by the interaction of the social and personal experiences of the visitor in a dialectical relationship with themselves, others and the precinct space'. Consideration of form, activity and meaning are useful, but insufficient, if we are to appreciate the factors that influence tourist experiences of urban spaces. In the tourist experience, it is the person and not the urban space that is the focus. The tourist experience involves interactions of the tourist with different aspects or elements of the space.

Experiencing place can be compared with meeting a person. For both, there are a series of qualitative components that mix spontaneously in a seemingly illogical manner.[41] The sense of place persists overtime despite some external changes just as the individuality and distinctiveness of a person persists through life.[42] Smaldone et al.[43] note how both people and urban spaces are unique and yet at the same time ever-changing. The comparison of the experience of place with the experience of meeting a person is interesting but has its limitations because, no matter how emotive it may be, experiencing place is primarily about inanimate buildings and features. Berleant[44] compares the more intense experience of place with an 'encounter with the noblest works of art, whose force overwhelms and engulfs those who engage with it'. These comparisons may be perceived to be ambiguous and yet they provide some useful insights in the nature of experience of place.

Sometimes people refer to the 'vitality' of an urban area or town. According to Montgomery,[45] this refers to (i) the number of people across different times of day and night (ii) the number of cultural events over the year and (iii) the presence of an active street life. Vitality of an urban space is dependent in part on the diversity of uses in the buildings around the space and within the urban space itself. Montgomery[46] also refers to 'transactions' these being exchanges of 'information, friendship, material goods, culture, knowledge, insight and skills, as well as the exchange of emotional and psychological support'. He argues that without 'transactions' spaces become lifeless and it is a wide range of uses that make these 'transactions' possible. To promote 'transactions' and to encourage vitality, the trend in urban planning and in urban regeneration projects is to encourage mixed-use developments.

## 8.2 HOW BUILDINGS AND SPACES ACQUIRE MEANING

The tourism experience of towns and cities takes place in a context of buildings, streets and urban spaces. To understand experience, the discussion cannot however be limited to a discussion of 'form'. A closer look to meaning is required, including how buildings and spaces acquire meaning. Space imbued with meaning becomes place[47] and hence it potentially becomes a location of interest to tourists.

Beyond functionality, architecture communicates meaning.[48,49] For example, the cathedral is the focal point and the dominant skyline feature of the medieval European town. This is indicative of the dominant role of religion in the lives of people in medieval times.[50] Different architectural features have different symbolic meanings (e.g. a spire suggests that the building is a church, a crenellated parapet evokes the idea of a fort or castle). The architectural symbol can also indicate an abstract idea or emotion (e.g. a portcullis is a symbol of strength and defence). The communication of architecture is dependent on a process of codification whereby people understand messages on the basis of pre-determined meanings.[51] The cultural environment, including buildings and spaces, constitutes a system of meanings structured around a complex amalgam of codes.[52]

Strike[53] notes that architecture can be read in different ways depending on the type of construction, the structural system, the details and the materials used. He describes how a building can be a symbol of a specific place. The best known example of this is the Eiffel Tower being an immediately recognisable symbol of Paris.

According to Grauman,[54]

> Members of a culture group invest places and people with meaning and significance. Not individuals but people agree on what is a forest or a jungle, what is downtown and what is suburb. It is essentially the language that people, that is the members of a language or cultural community, share (with, of course, individual, sometimes idiosyncratic, variations) that communicates meaning.

The acquisition of meaning of buildings and spaces takes on a particular significance with reference to historic areas. Meaning is intrinsic to historic areas – it is the history of the place which is the meaning and which, in turn, could be a source for other meanings which the visitor will derive based on one's background. Place acquires its distinctive meaning through the interaction of human sensibility and the material environment.[55,56] For Sancar and Severcan,[57] the meanings ascribed to specific environments are products of interactive processes involving the individual, the setting and the broader social world. Urban context often becomes a repository of social and cultural significance, as well as the embodiment of a collective memory.[58] Gospodini[59] refers to historic urban cores as fragments of the city that have become rich in meaning and that are subject to different interpretations depending on the contexts.

The meaning of buildings, spaces and artefacts are created through the person-object interactions that are heavily influenced by the tourist's own cultural and social background.[60,61] The heritage experience entails coding systems which are not stable nor are they universal. It can be compared to a language for which each individual possesses their own personal dictionary which is constantly changing.[62] For heritage artefacts, the visitor often manipulates the artefact on offer, bringing one's own emotional and intellectual responses. This results in meanings which may or may not have been previously conceived.[63]

According to Voase,[64] meaning does not reside in externalities but is defined in the mind of the observer at the point of observation. He illustrates this by referring to research carried out at Madame Tussuads in London during the 1980s. It was discovered that the satisfaction of visitors was not derived from the artistry and the likeness of the wax figure. Satisfaction resided in the exploration of the memories, feelings and emotions which the visitor associated with the person represented by the wax figure. As a result the layout and presentation of Madame Tussuads was redesigned. Rather than in roped-off areas, the wax figures were placed into settings where visitors could walk around them, touch them and simulate interaction represented by the figure. To some extent, visitors relate to historic buildings and sites in a similar manner in that the meaning as interpreted by the visitor plays an important role in the experience.

Meanings are dependent on the events and artefacts which the buildings and spaces are seen as representing. Visitors produce 'their own subjective experiences through their imaginations and emotions' and 'imbue objects in the setting provided with their own personal meanings'.[65] It is meaningful social experiences, rather than simply the spatial attributes of physical settings, that drive attachment to places.[66] Timothy[67] notes different levels of meaning depending on the extent of personal attachment, including personal, local, national and world. There are instances where an artefact, viewed as world or national heritage by one person, is considered as personal by another. In considering meaning of place, it is useful to distinguish between people who are very familiar with a place (i.e. residents or regular visitors) and those who are much less familiar (i.e. tourists). The way the place will be read and interpreted will be different between these two groups of people.

## 8.3 TOURIST EXPERIENCE AND THE SEARCH FOR MEANING

The search for meaning is described by some writers as a primary motivation in a person's life.[68] Its relevance to the tourism experience should not be underestimated. Leisure is not just a search for pleasure but also a search for meaning.[69] Hannabus[70] argues that tourists do not just search for what is different from their everyday lives but are 'also in search of "meaning", of the "genuine" or "authentic" holiday'. In academic literature, reference is sometimes made to 'existential authenticity' with some tourists feeling themselves more authentic and more freely self-expressed than in everyday life because they are 'engaging in non-ordinary activities, free from the constraints of daily commitments'.[71] These may entail feelings and emotions that have little or no reference to context. It is about being in an environment that enables you to be at peace with yourself. The meanings of buildings, spaces and artefacts are heavily influenced by the tourist's own cultural and social background as well as by memories, interests and concerns.[72,73]

Cohen[74] develops a typology of tourist experiences and distinguishes between five modes of tourist experiences. He presents them in an ascending order from the most 'superficial' motivated by 'pleasure' to one most 'profound' motivated by the search for meaning. Cohen recognises that the tourist may experience different modes on a single trip but presents them separately for analytic purposes. Broadly speaking, the five modes can be categorised into two; those for which the site does not involve meaning (namely 'recreational' and 'diversionary') and those for which meaning has a pivotal role in the tourist's experience (namely 'experiential', 'experimental' and 'existential'). In the 'recreational' mode, the trip restores the tourists' 'physical and mental powers and endows him with a general sense of well-being'. Although the experience may be interesting for the tourist, it will not be personally significant. The 'diversionary' mode is 'a mere escape from the boredom and meaninglessness of routine, everyday existence, into the forgetfulness of a vacation'. For Cohen, the 'experiential' mode involves the quest for meaning outside the confines of one's own society. It is to some degree spurred by alienation and a sense of lack of authenticity at home. Cohen draws comparisons between 'experiential' mode and a religious pilgrimage in that both involve a search for what is perceived to be authentic. In tourism, however, the authenticity will not provide a new meaning and guidance, even if it may reassure and uplift the tourist. 'Experimental' mode involves an effort to rediscover oneself in another context because of the alienation which deeply affects the individual. 'Existential' mode involves the search for a better world elsewhere resulting from a feeling of living in the wrong place and at the wrong time.[75] Cohen's five modes can be categorised into two broad categories: those that involve meaning and those that do not. This is also reflected in McIntosh's[76] distinction between 'mindful' and 'mindless' state of mind in relation to the heritage visitor experience. 'Mindful' refers to visitors being 'sensitive to the context, actively processing information and questioning what is going on in a setting', resulting in greater appreciation and understanding of the heritage and its meaning. In contrast, little or no understanding will result from a 'mindless' tourism experience and the visitor will not take note of the meaning that may reside within the site or feature being visited.

## 8.4 THE ROLE OF NARRATIVE AND THE PAST IN THE TOURIST EXPERIENCE

Another parallel that can be drawn is between the 'meaning' component of sense of place and narratives. The social identity of people is constructed and sustained by means of webs of stories and narratives produced by people.[77] According to Jamal and Hollinshead,[78] 'There are no stories waiting to be told and no certain truths waiting to be recorded; there are only stories yet to be constructed'. Narratives are 'constructed' by people and they may, or may not, be based on historical fact. More importantly, heritage narratives are stories that people tell about themselves, about others and about the past.

A narrative is better appreciated by tourists if there is a heritage site which, as Rickly Boyd[79] describes, provides 'the material and the setting to combine lived experience with myth in the production of a uniquely personal tourist narrative'. Narrative may further support and reinforce the tourist's experience of place. This is all the more evident in heritage sites where stories reinforce a place's uniqueness or focuses on narratives of national significance.[80] A successful tourism product is an interpretation of the local history or narrative within the context of the historical experience of the tourist or visitor.[81]

For some tourists, part of the experience involves understanding 'how life really was'. Life conditions that might be unthinkable in the present day can be 'experienced', or more fully understood and appreciated, through the narrative.[82]

Ashworth[83] states that heritage-related tourism manifests itself in a mix of preserved buildings, conserved cityscapes and morphological patterns, as well as place association with historical events. These are the resources that create the tourist-historic city. He describes the tourist-historic city as a cluster of heritage tourism sites and facilities and also 'a more holistic idea of the heritage city as a place where tourism activities (both heritage-related and non-heritage related) occur'.[84] An argument relevant to historic areas is that made by Relph[85] when he states, 'The essence of place lies . . . in the experience of an "inside" that is distinct from the "outside" '. In the historic environment, there are competing demands and underlying tension between past and present cultures and between the familiarity of the old and the notion of progress attached with the new.

Very often the tourist experience of a historic area is mostly about how the tourist relates to its past. The relationship between the tourist and a historic area's past manifests itself in a number of different, albeit interrelated, ways.[86] From features and the historic ambience, tourists can tell that the place is historic without necessarily getting to know more about the place. Some tourists enjoy being immersed in a historic context, even if they know very little about the history of the place. In some historic spaces, there is a total absence of modern-looking structures or buildings and this gives the sense of going back in time. In spaces where there are no evident modern features, the tourist experience is not just about being in a different place but also possibly the feeling of being in a different time. The feeling of being in a different time is further reinforced in spaces where there are no cars, like in Valletta's stepped streets.[87] For the more imaginative tourist, there might also be an element of visualising past life or past events. For example, one interviewee in research on Valletta[88] spoke about how she visualised life in Valletta hundreds of years ago, with knights entering and exiting the various baroque palaces of the city.

Urban heritage enables the visitor to see tangible evidence of narratives of the past, thereby making them more real and the tourist experience more interesting and enjoyable. The urban fabrics of most historic areas convey a range of narratives, some associated with the local identity, whereas others are linked to past foreign occupiers. The appearance, form, detail and material of most buildings show that these were built many years ago and therefore they have a story to tell.[89]

To better understand urban quality, it is useful to contrast historic urban areas with modern spaces and development. Maitland and Smith[90] note the lack of atmosphere and tourism potential in postwar new towns. Norberg-Schultz[91] argues that the modern city does not offer enough possibilities for living as its streets and squares are no longer places for people, but simply a means of communication. Similarly, Berleant[92] notes that many live in anonymous bland environments. He describes it as living 'in industrialised landscapes, in insular factories, strip malls and office towers, moving with clockwork regularity along highways that are self-propelled conveyor belts to faceless apartment buildings and generic suburbs'. Negative qualities commonly attributed to modern development give some indication as to the positive qualities or attributes that historic urban areas potentially offer including variety, interest and authenticity. These observations are relevant to some parts of the modern city, but not all of it as more and more cities give greater attention to the qualities of the physical environment, especially urban spaces.

For many tourists, history and narrative are dominant in the experience of a historic area. It may be that narratives of place are not grounded in historical fact, but for most tourist sites, it is the idea of 'history' and their position within it that provides the narrative and the meaning that visitors will find interesting. Many tourism experiences are dependent on the availability and communication of narratives. Some places are part of the tourism itinerary because they are associated with powerful stories.[93] Ashworth and Tunbridge[94] explain how two major historic episodes in Malta's history, namely the Great

Siege of 1565 and Malta's role in World War II (1940–1944), provide a narrative for 'almost an ideal tourism product'. The narrative features and qualities that make for an 'ideal' product include easily understood dramatic events with few historical ambiguities. Moreover, the narrative is capable of interpretation through experiences, with the sites, buildings and artefacts lending themselves to easy identification by the tourist.[95]

## 8.5 TOURISTS AS PERFORMERS

The 'tourist gaze' concept developed by Urry[96] has been very influential in portraying the tourist experience as a visual experience; 'the tourist gaze' suggests that people travel to destinations that are visually striking. Ek et al.[97] argue that tourism requires new metaphors based more on being, doing, touching and seeing rather than just 'seeing'. This turn to performance in tourism studies is a reaction against the 'tourist gaze' and other representational approaches that privilege the eye.[98] A metaphor that is being referred to frequently in tourism literature likens tourists to performers who 'not only consume experiences but also co-produce, co-design and co-exhibit them'.[99] Edensor[100] notes that social life is performative and dramaturgical, built of roles according to social contexts. He contends that tourist performances can be considered as a continuity of the enactments in non-tourist settings. Tourist performances are shaped by the constraints and opportunities that tour structures and tourism destinations produce and are informed by the symbolic meanings and spatial organisation of tourist destinations and sites.

Ek et al.[101] contend that the tourist is not a passive sightseer consuming sites in prescribed fashions. In the act of consuming, tourists turn themselves into producers and the distinction between producer and consumer is dissolved. Tourists enact and inscribe places with their own 'stories' and follow their own paths. There is always an element of unpredictability. The places and performances that tourists enact are never completely identical to the scripts in marketing material, guidebooks and so on. Tourists experience places in ways that involve senses other than sight, that is, touch, smell and hearing. 'Tourists encounter cities and landscapes through corporeal proximity as well as distanced contemplation'.[102] Similarly, Wearing et al.[103] highlight the role of individual tourists themselves in the active construction of the tourist experience.

There are different types of stages or tourism spaces catering for different tourism activities and ventures.[104,105] An enclavic tourist space has clearly defined boundaries and a high level of stage management of the tourism performance as well as in-house recreational facilities, normally including displays of local culture. Tourists are characteristically cut off from social contact with the local people and are shielded from potentially offensive sights, sounds, and smells.[106] By contrast, a heterogeneous tourist space is a multi-purpose space in which a wide range of activities and people co-exist. Tourist facilities coincide with commercial activity and domestic housing. Tourists mingle with locals in a space with blurred boundaries.

In a heritage area, tourists can wander around and explore and also come in contact with the activities of locals. It can be considered therefore as a heterogeneous tourist space that allows ample flexibility for tourists' performances.

Typologies are useful in trying to understand the behaviour of different groups of tourists.[107] The discussion on typologies is not about different types of tourists but about describing different types of tourist practices or 'performances'.[108] Griffin and Hayllar[109] propose a typology on the basis of which they highlight how a tourism precinct can be experienced in different ways by different people provided that the 'precinct offers opportunities for different "layers" of experience'. The more 'layers of experience' offered by an urban space, the more opportunities for leisure activities are offered and hence the more likely will it be enjoyed by tourists. They derive the visitor typology from a study of waterfront precincts in urban areas in three Australian cities. Those visitors who want to move beyond the façade of a precinct to find their own way can be described as 'explorers'. They seek to discover the complexities and qualities of the precinct. Several visitors to one of the areas commented on how they enjoyed exploring behind the buildings and through the alleyways. The 'browsers' are content to stay within the confines of the main precinct area and to follow the tourist routes. The 'browser's' visit involves walking along a pedestrian thoroughfare containing places of historic interest and some restaurants and shops. The experience of 'browsers' does not have the depth of the 'explorer', but they are interested in capturing the 'experiential breadth' of the precinct. On the other hand, there are visitors who visit a precinct as just another stop on their itinerary of a city's attractions. These can be described as 'samplers' and are often concerned purely with visiting a specific attraction rather than experiencing the precinct for its own sake. For 'samplers', the urban precinct is a place that, over a short span of time, gives some idea of the area's history but more important, it is the place where souvenirs can be bought. Alternatively, samplers use the area as a place of refuge from the more dynamic environment of the nearby metropolis.

The conceptualisation of the tourist has shifted from that of an 'itinerant gazer' to that of a person interacting with the surroundings.[110] Orbasli[111] describes the experience of a visitor in an urban environment as one of 'discovery and appreciation' and refers to tourists wandering, lingering and taking in the surroundings. In a study of tourism to Islington and Bankside in London, it was noted that visitors were drawn by the qualities of place rather than specific attractions. Visitors enjoyed the 'broader qualities of place – the physical environment created by architecture, building, streetscape and physical form, combined with socio-cultural attributes such as atmosphere and being in an area perceived as "cosmopolitan" and "not touristy" '.[112] It is the uniqueness of a place, rather than its more general qualities that makes the sense of place of the tourism destination.[113]

In a study of two waterfront precincts in Australian cities, Griffin and Hayllar[114] noted that one of the activities reported was simply wandering around. For some, this entailed 'a sense of exploration, with visitors mentioning the notions of "getting lost", "getting one's bearings" '. In some cases, there was 'a belief that the wandering would lead to discoveries that engaged the visitor'. On the other hand, Burns[115] argues that whereas 'people enjoy "crooked streets" and the richness of urban experience', they are most afraid of being lost and disoriented.

'Mystery' is a term sometimes used with reference to places with 'character'. Kyle et al.[116] note that 'for mystery to be present, there must be a promise of further information if one could walk deeper into the scene'. Examples of 'mystery' include 'the trail that bends, the road that turns, or the vista temporarily hidden from view but accessible through a simple shift in position'.

*Chapter 9*

# Urban Conservation and the Tourist-Historic City

## 9.1 URBAN HERITAGE AND TOURISM

No urban area is the product of a single historic period. Throughout history, cities were established and planned. They grew and expanded. They were often attacked or invaded and ultimately, they were replanned and rebuilt. In the process, they acquired a collective identity of cultures, architectural style and human life.[1] The historic city, and hence urban heritage, is 'essentially a product of the time and place of those who have created it and continue to shape it'.[2] Heritage is often the subject of debate among conservation practitioners and academics. In spite of its widespread use, or maybe because of it, there isn't one common understanding of the term. Millar[3] claims that 'heritage is about a special sense of belonging and of continuity that is different for each person', whereas Graham et al.[4] refer to heritage as 'the contemporary use of the past'. According to Tunbridge,[5] 'heritage usually signifies tangible or intangible features, particularly of the built environment or cultural lifestyles respectively, which are deemed to be valuable inheritances of the past'.

An alternative way of thinking about heritage is that it is not a matter of historical resources as such, but rather of the meanings that are derived from them.[6] In this context, the discussion in the previous chapter on meaning and its relevance to the tourist experience is especially pertinent. In this chapter, the tangible aspect of heritage is the focus of the discussion and hence the use of the term 'urban heritage'; to emphasise the built environment context.

There is an interactive process between the present and the past. People perceive the past in the light of the present and vice versa. Heritage sites are interpreted by individuals and by communities on the basis of their present experience. Vahtikari[7] explains:

The gaze of heritage is directed towards the past but is always interpreted from the current perspective, for present and future purposes, and is infused with the concerns and the uses of the present.

The past is valued and understood differently depending on the background and past experience of the interpreters. Because of cultural diversity and differences in experience, different histories will perceive different things as significant. Not all cultures will share the same concepts of what constitutes heritage. A heritage site is a complex network of meanings. It is a cultural mix formed in the past with meaning which transcends the time of its origin. Such sites have value as they are saying something to present society. Heritage sites operate at two different levels. On one level – as material things – they are a historical record. At another level, they convey ideas, stories and values.[8]

'Value' is central to the concept of heritage. Society will not expend time and resources on the protection and preservation of an artefact or feature unless it is considered of value. Heritage sites are identified, conserved, managed and interpreted according to the cultural values they are seen to represent. Articulating heritage values is problematic as several diverse values exist, in coexistence or in competition with each other. The challenge is made more difficult as values change in different cultural contexts and change also over time. The value of a cultural object is not intrinsic to it but is something that is attributed to it by society. Heritage may be seen as a continuous cultural process in which meanings and values are created, negotiated and transmitted.[9]

Most cities have an area which is historic and from which the city would have originated and developed. In many cities, the historic core normally plays a crucial role in tourism as the location of the more important attractions. The historic core is often an attraction in its own right with many tourists spending time walking and exploring the area.[10] Tourism motivated by heritage is driven by historical resources that are anchored in the built environment and in other tangible artefacts. Normally, it also refers to intangible attributes and associations, including narratives. So pervasive has such heritage tourism become that for the past three or four decades, the 'tourist-historic city' was recognised not as the exception but as the norm of urban centres in much of the world, as they increasingly vied with one another to offer their heritage to tourists.[11] A city's urban heritage is often a potent tool to attract attention both nationally and internationally. For the more popular tourist-historic cities, the problem is not so much in attracting tourists. The challenge for city authorities is to balance tourism, conservation and the needs of local residents.[12]

Not all tourism is concerned with historic resources nor are such resources invariably concerned with tourism. A discussion on tourist-historic cities is justified because tourism in its various forms has a critical role in the development of historic resources. Moreover, historical resources form an equally critical part of a growing tourism industry. Interestingly, the symbiosis of the two has become a major activity of cities and a major force in the way that modern cities have evolved.[13]

The polluting factories created in cities in the nineteenth century were moved out or closed down altogether. City authorities sought newer and cleaner economic sectors to replace them. Tourism often was the preferred option. Urban heritage is a popular tourism product and therefore many cities reconsidered or reinterpreted their urban heritage as a marketable attraction. Even places previously perceived as centres of industry began to develop new heritage images.[14] For example, Bradford, England, was a thriving trade and manufacturing centre in the nineteenth century. Industrial decline across the UK left Bradford with derelict warehouse buildings, soaring unemployment and a sizeable Asian immigrant community. At the end of the past millennium, city authorities were keen to promote Bradford as a historic town with Asian attractions. They embarked upon a range of initiatives to make the city more attractive, including, in 1987, the staging of the Bradford Arts Festival in a newly refurbished warehouse district.[15]

For historic towns faced with limited financial, heritage tourism is seen as a significant economic alternative. Today historic towns and quarters are competing to attract tourism, and previously unknown locations are appearing on the heritage market. So intense has competition between tourist-historic cities become that they increasingly need to look beyond a fixed heritage offer and rework their historical resources to generate new heritages.[16] For example, they may develop new heritage attractions by establishing appropriate visitor facilities with existing historical resources. They may revamp existing heritage attractions so that they provide a more interactive experience for visitors or develop heritage-themed events grounded in the city's history and narratives. Historic areas in a city provide added value and enhances the tourism potential of a city destination. A new urban society has emerged, seeking leisure, culture and a high-quality environment, and cities have moved from being industrial centres of production to becoming centres of consumption.[17]

## 9.2 THE CONSERVATION OF HISTORIC AREAS

In World War II, many elements of the architectural and urban heritage of the European cities were destroyed. In the immediate aftermath, different practices were applied for the reconstruction of the demolished cities' parts. In some cities such as London, Berlin and Rotterdam, war-damaged urban areas were cleared of remaining buildings and new modern-style buildings were constructed in their place.

This approach to conservation was continued in many European countries up until the 1950s and 1960s. It was based on the principle that new buildings should abide by present-day norms of design and construction. The approach was to create modern town centres conveniently laid out to meet land use and traffic requirements unencumbered by the legacy of the past with the intention of creating urban environments that were perceived to be healthier. Road networks were reorganised to facilitate car use. It was a time when old buildings were often regarded by public authorities as an obstacle rather than an aid to regeneration. Except for buildings that were of outstanding historic value, there was little interest in conservation as the urgent needs for new housing were better served by new construction. In later years, post-war remodelling of city centres and the new urban forms were heavily criticised because of their lack of identity and their placelessness. An alternative approach to post-war reconstruction was adopted by Warsaw, Dresden and other cities. These cities chose to faithfully rebuild their war-damaged city centres exactly as they were before the war, adhering to the urban forms of the past. The basic principle of reconstruction here was that the city's identity and culture would be revived if the city was rebuilt as it was.

As from the latter part of the twentieth century, societies gave greater value to the symbolic and artistic significance of historic buildings. Historic areas thus acquired a dual nature – their functional use and their heritage value. New approaches to urban conservation gained ground with growing awareness of the need to ensure sustainability of the built environments. The focus shifted to the sanctity of authentic historic fabric and the custodianship of buildings for future generations.[18] Moreover, urban conservation prioritises the improvement of residents' quality of life and the support of economic development. The latter in particular requires greater attention to making historic urban areas more attractive for both residents and tourists.[19] Two approaches to conservation developed. The first is based on the tradition of planning and managing the city as a built form of material culture, while the second approach is based on the meaning of place. Common to both approaches is the search for a new integration of tradition and modernity, of the built and the non-built environment.[20]

In many towns and cities, the central areas are the most vulnerable. Lack of investment in building maintenance and renewal result in a process of gradual decay and eventual dereliction. In many cases, the central areas are historical and part of the city's identity, so a process of decline is detrimental to the town's urban heritage. This vulnerability occurs directly through neglect and abandonment of the central areas or indirectly because of relocation of central area for residential and commercial uses to other parts of the town.

At the greatest risk of decline are the central areas of towns and cities that are away from the tourist circuit. For some towns, this is the typical chicken and egg situation. What comes first the renewal of the historic centre or the income that comes from tourism? Without a renewed historic centre, tourists will not come and with no tourists the town authorities will not have the funds to renew the historic centre.

Built monuments have traditionally been mainly defined as individual buildings and their conservation considered merely as actions directed at the monument's structure. The concepts of 'a monument' and its 'conservation' has changed over time. Today it is not just the individual historic building that is valued but also the cultural landscape and the human activity. Conservation goes beyond the structure and fabric of the individual building. Conserving a historical milieu requires an interactive balance between the different factors involved in the monument including the setting (buildings, structures and landscapes) and the users (residents, visitors and other users).[21] Orbasli[22] draws a distinction between urban conservation and building conservation. The latter refers to the conservation of an individual building or groups of buildings. On the other hand, urban conservation is not limited to the renewal of the physical fabric of buildings. It also involves the urban pattern, streets, open spaces, green areas and urban vistas. Moreover, it requires the services of a much wider range of disciplines. It is affected by political decision-making at local and national levels. There is also the social aspect, with residents, property owners, business interests and other users all being part of the conservation process.

Ashworth and Tunbridge[23] developed the concept of tourist-historic cities to characterise areas where historic urban structure and architecture combine to create a place-based heritage product to be consumed by visitors and tourists. They drew attention to both the benefit and the problems of tourism in historic cities, and how planning and management might foster a mutually beneficial relationship.[24,25] They noted that approaches to historic city management evolved from one of preservation (dating from the nineteenth century), to conservation planning (from around the 1960s) and heritage planning (from the 1980s). These approaches reflect different objectives and values. In preservation, for example, the main goal is ensuring the survival of the building. Heritage planning, on the other hand, focuses on the consumption of the area as heritage, within an overall context of conservation of the historic urban area. Ashworth and Tunbridge[26] argue, however, that there is no requirement for one approach to replace another, so during the last part of the twentieth century, heritage planning co-existed with a continued concern for conservation and preservation.[27]

Urban areas are dynamic and are subject to change. In historic areas, the urban conservation processes seek to direct change to ensure the appropriate maintenance and management of a historic area for the purpose of sustaining and enhancing its significance. Urban conservation is a long-term process over time that requires political, economic and social commitment to an area with the intention of providing better quality of life for its users. It cannot be considered in a vacuum. The process of conservation is closely linked to the overall context within which it takes place.[28] Conservation encompasses not only the physical urban fabric but also an understanding of spatial morphology. It also considers the social dimension and community values that make the urban heritage distinct from the more intrinsic physical qualities of the singular built heritage. Both the conservation and the effective management of the historic townscape are dependent on a sound understanding of its historical and spatial structures, an understanding of social and community structure, and an appreciation of the private and religious uses of urban space.[29]

Bandarin and Ron van Oers[30] argue that the spatial form of a historic area necessitates adherence to the existing traditional morphology and avoidance of drastic changes and interruptions in the continuity of the urban fabric. Effective conservation also requires that, as far as is practicable, the traditional uses of the area should continue functioning as they constitute elements of a historic area's identity. They advocate that, as much as possible, the trend towards gentrification needs to be controlled to avoid the loss of the historic area's identity. Careful planning should seek to achieve a better quality of life for residents by reinforcing the residential function while concurrently increasing the area's attractiveness to visitors to contribute to economic growth. Ebejer and Dimelli[31] contend that historic areas must be considered in their totality. Their conservation should seek to develop strategies and plans that identify and protect the elements that contribute to the values of the historic towns, as well as the components that enrich and demonstrate their uniqueness. The conservation of historic areas should seek to promote synergies between stakeholders, institutions and residents so as to develop a sense of ownership of the urban heritage by all who have an interest.

Conservation and regeneration are often presented as largely complementary processes.[32] Urban regeneration projects often include the conservation of historic buildings and their adaptation to a modern-day use – one for which they were not originally intended. The inclusion of conserved historic buildings in an urban regeneration scheme is perceived to add quality and place distinctiveness.[33]

Although often perceived in synergy with each other, there are often underlying tensions between conservation and regeneration. First, there are different views to what constitutes valid conservation. Conservationists have a more purist view and would argue that major alterations and demolition are not acceptable in any circumstance. It is not uncommon for conservationists to be made deeply uncomfortable by transformations of historic buildings in regeneration schemes. On the other hand, there are those who argue that demolition and major alterations of historic buildings should be actively considered if it will improve the feasibility of an urban regeneration scheme. Ultimately, the long-term viability of the historic buildings is dependent on having regeneration that successfully installs new life into an urban area. A second source of tension between conservation and regeneration is the scepticism in the property sector over the importance placed upon conservation policy objectives.[34] Often, pro-development lobbyists put pressure on legislators and urban planners towards a planning system that is more amenable to interventions on historic buildings.

Why conserve? There are many reasons for the conservation of buildings and historic areas. The main ones are intellectual and psychological. The intellectual justification for conservation refers to the retention of historical buildings because of their artistic architectural or historical qualities and therefore for their role in highlighting the cultural achievements of a society.[35] In the earlier years of urban conservation, it was the intellectual rationale that made societies conserve particular structures such as monuments, important state buildings, religious structures and the houses of the ruling elite. These were retained because they were seen to embody important cultural traditions of a society and nation. It is because of the intellectual rationale that the role of the state in urban conservation has developed.

The psychological rationale refers to people's reactions to the increasing pace and scale of urban change. Throughout the twentieth century and across many cities, the loss of individual buildings as well as large urban areas through redevelopment made residents feel a sense of loss and dislocation. Towards the end of the twentieth century, it was a feeling of loss that drove many community groups to campaign for the retention and conservation of historic buildings. A wider popular involvement in urban conservation has served to widen its scope beyond the designation and protection of a few important structures to more extensive protection of areas of cities, including older buildings that may be considered as ordinary and mundane. This brought about moves towards the protection of larger parts of the city and not just individual structures. It has also broadened the scope of conservation beyond the protection of historic gems to include some twentieth-century urban landscapes.[36]

Many city residents become very familiar with the historic areas of their city and hence grow a strong attachment to them. Many would associate them with their younger days or important events in their lives. People attach considerable value to aspects of their immediate environment that give them a sense of identity and pride of place. Conservation and the upgrading of historic urban spaces contribute towards making them places for public congregation and activity. Urban conservation re-affirm residents' feelings of identity and sense of belonging while reinforcing the identity and image of the city. The historic city is seen as a tangible link between past, present and future. Urban conservation reduces the disorientation that may result by the replacement of familiar buildings with new and unfamiliar urban forms. It is often claimed in academic literature that familiarity with places is valuable in maintaining an individual's psychological stability and excessively rapid environmental change may upset this stability.[37] Lynch (1960) argues that 'the excitement of the future should be anchored in the security of the past'. Historic areas provide a sense of place in a manner that areas of modern development are less likely to provide. As discussed in chapter 8, the sense of place emanates from the meanings associated with buildings and the urban spaces. The collective memory of a community is dominantly place-bound and is expressed through the physical attributes of places. Rapid change in these physical features causes loss of the collective memory that may result in some level of social disorientation.[38] Another reason for conservation is the intrinsic aesthetic, cultural or historical value of the historic building or group of buildings. Conservation of historic areas contributes largely towards upgrading environmental quality, thus serving as a fundamental catalyst for sustainably managed change.

The economic justification for urban conservation has several dimensions. Urban heritage can be used in place marketing initiative to influence the perceptions of a place as a suitable location for investment, residence or recreation.[39] The conserved historic city may contribute to a high-quality environment that is attractive to potential investors and new residents. Above all, attractive historic areas are a vital part of a city's tourism product as it generates heritage tourism and also provides additional activities and places to visit for all visitors to the city.

It is not uncommon for buildings to become redundant, unable to serve the purpose for which they were originally built. Some are adapted to new uses but others remain vacant. For their cultural and townscape value to be appreciated, the historical building needs to be brought back into viable use. Tourism enables reuse and encourages new economic activity, as some tourism uses are flexible and could be adapted into the difficult internal spaces of historic buildings. The adaptive reuse of existing resources reduces new construction and related environmental pressures, while increases use in central locations.[40]

## 9.3 FUNDING AND THE ROLE OF PUBLIC AUTHORITIES IN URBAN CONSERVATION

The roles of public authorities in urban conservation are many and diverse. First of all, there is the regulatory role. Regulatory functions relating to urban conservation are often an integral part of the urban planning process at the levels of forward planning and development control, both of which are discussed in section 2.4. Within a city authority, there is often an agency or a department that is responsible for the drafting of regulations, guidelines and policy documents relating to urban development in general and very often also relating specifically to urban heritage and historic areas. To be effective, regulations need to be enforced, whereas policy documents need to be applied. It is the responsibility of the same agency or department to ensure that regulations are adhered to.

It is also the role of the public authority to apply guidelines and policies, normally through the development control process. Development control is that process by which property owners are required to acquire a development permit before being allowed to go ahead with a development on their property. They are required to submit an application for development to the relevant authority, which, in turn, is obliged to process the application and assess it with reference to the applicable guidelines and policies. The application for development may be refused but it is normally approved, sometimes following discussions between the applicant and the public officer. Very often alterations to what is being proposed are negotiated between the two sides.

Public authorities operate the development process for the entire city, including historic areas. In the case of historic areas, the process is normally more rigorous as inappropriate developments could impact not just the street where they are located but also the entire historic area. Urban conservation is regulated by means of legislation which in many countries is an integral part of urban planning legislation.

An urban planning tool that is widely used across Europe is the formal designation of historic areas as 'Urban Conservation Areas' or similar. Designation signifies that these areas are subject to rules and regulations that protect the historic fabric and permit limited and rigorously controlled modern intervention.[41] Another vital planning tool is the 'listing' of buildings because of their historical or cultural value. The term is derived from the act of including the building in a list of properties that are of value, be it historical, cultural, archaeological or for some other reason. In some jurisdictions, this is referred to as 'designation' of the property. The concept of using legal means to protect historic buildings developed over many years and involved campaigning and lobbying with politicians in Britain and elsewhere.[42] At the very least, designation provides protection from damage but it does not guarantee that the owner provides the necessary funds to maintain the building. Of itself, designation turns a property into a monument laden with ascribed values. This changes the way it is perceived, even if nothing is actively done on the property.[43] The mindset and attitude of owners, users and the public at large will shift as designation raises awareness of the property's cultural and historical values.

Designation is not however a total guarantee against the loss of the building. Some protected buildings are lost to natural misfortune, wrong planning decisions or commercial greed. According to Ashworth and Tunbridge,[44] losses amount to 2 to 5 per cent annually of the designated urban heritage properties but this is dependent on the jurisdiction and the predominant public and political perception of heritage at a particular time. Moreover, buildings have a physical life span that cannot be extended indefinitely. They will fall into dereliction, and eventually ruin, unless they are properly maintained. Also important is the visual damage to townscapes and historic ambience caused by insensitive new development. Paradoxically, over-use and over-commercialisation of popular tourist sites can also be a threat to designated historic sites.[45]

Urban heritage in historic towns often acts as a primary attraction for tourists. They are a free commodity in that tourists and locals can enjoy their visual and symbolic value without having to pay for them.

Hospitality services and commercial outlets within the town are secondary attractions but they constitute the greatest opportunity for financial gain. Although tourism is not a direct financial resource for conservation, it indirectly opens up previously unavailable finances for investment. Tourism is potentially an important catalyst for the safeguarding of historic fabric and the initiation of conservation on an urban scale. Appreciation of the historic environment by visitors not only becomes a reason for conservation but can increase local interest in the urban heritage.

Before the 1990s, the restoration of historic monuments and urban conservation was financed mostly by governments and public authorities. More recently, governments are less able to dedicate financial resources to urban conservation. Limited financial resources are a major constraint for the conservation of urban heritage, with governments often handling other more pressing priorities normally related to economic development and social protection.[46]

The investment in infrastructure and urban conservation does not generate direct financial return to the public authority, but it does generate benefits to the local community through knock-on effects and the creation of jobs.[47] Apart from urban improvements, developing tourism in a historic area may necessitate investments in infrastructure. For example, a new transport interchange is a costly investment, with no direct return to the authority that funds it. It is however potentially a long-term investment providing better mobility and improved communication. Improved transport infrastructure supports the conservation of a historic area as it benefits residents and businesses as well as tourism activity. The conservation of a historic area can potentially increase the economic activity and hence generate increased revenue to the city authorities; revenue that, in turn, can be re-invested in conservation.

In any development, it would be preferable if the income generated from the project would be sufficient to sustain both the capital cost and the running cost. For historic buildings, achieving financial sustainability is difficult because of the high cost of restoration and the difficulty to adapt a historic building to a viable use. This is why funding from national governments and international agencies for such projects is important. In an EU context, the use of European Regional Development Funds (ERDF) is essential to carry out conservation projects which otherwise would not have been possible from national funds.

Private-sector financing is increasingly essential for development and for urban conservation. This, however, results in increased private-sector influence and may in some instance challenge the public authority's role in development control. Urban planning objectives derived by public authorities may be seen as limiting competition or discouraging investment. This puts pressure on city authorities to compromise on sustainable development objectives.

In some instances, public funds are used to leverage investment from the private sector. One example of this was a funding scheme for the restoration of balconies in Valletta. Timber balconies are a characteristic feature of Valletta and other historic towns in Malta. There has been a gradual loss over time of these balconies, either because of decay or because of reconstruction with inappropriate materials. As a means for safeguarding this urban feature, public authorities in Malta offered to finance a substantial part of the cost of timber balcony restoration, with the remaining part being financed by the property owner.

The use of church buildings and monasteries for religious purposes has declined in recent decades. In past time, peoples' religious fervour had produced architecturally outstanding buildings often located at or close to the town centre. In modern times, these building have become an important part of the city's tourism itinerary. The cost of maintaining these buildings is significant more so when a major conservation project is required. Church buildings have often resorted to cultural use to generate funds to ensure that the historical and architectural values of these buildings are safeguarded. Church buildings were designed with both 'audience' and 'stage' facilities and therefore use for concerts and conferences seems appropriate,[48] even if for the former, the seating layouts and the acoustics are often less than ideal. Another potential use is exhibition spaces.

Financing the conservation of historic fortifications and town walls is particularly challenging. Networks of fortifications often have internal spaces within them. These are normally small and the layout is not conducive to efficient circulation.[49] Restoration is expensive. Identifying a use that generates enough income to cover capital costs is next to impossible and therefore reliance of public funds is inevitable. The same can be said for historic buildings albeit to a lesser degree. For historic buildings, the most obvious tourism-related use would be that of a visitor attraction or a museum, although other uses could be considered. Whatever the use, some alterations to the historic building would be required but urban conservations requirements necessitates that these are carried out in a manner that does not compromise the historic and cultural value of the building. It is often problematic or even controversial to strike the right balance between adapting to a new use and safeguarding the historic authenticity. The situation is made even more difficult because of mandatory requirements relating to fire safety and access for people with mobility difficulties.

Malta offers interesting examples on the restoration and adaptive reuse of historic buildings, including fortifications. For many years, Malta had difficulties to finance much-needed restoration projects. The urban heritage that was in public ownership was decaying and parts of it were becoming derelict. Towards the end of the twentieth century, there were some effort in restoring some historic buildings but these efforts were limited due to a lack of financial resources. Public investment in urban conservation picked up momentum following Malta's entry into the European Union in 2004. The most intensive period of project implementation was between 2006 and 2012, with virtually all projects being completed by 2015. Many properties in public ownership were restored and brought back into use. Most projects were made possible with significant EU funds under the ERDF Programme. These funds covered the capital costs but these were allocated on the strict proviso that the running of the newly established uses will generate enough funds to cover the running and maintenance costs. The objective of urban heritage projects was to restore and bring back to life historic buildings and spaces. In the case of fortifications, the projects converted what were previously war structures into places for tourism, leisure and the appreciation of heritage.[50]

## Case Study UC1: Restoration and adaptive reuse of fortifications in Malta

Malta possesses extensive fortification lines.[51] These also include fortified structures, such as forts, cavaliers and other structures. The military design of defensive fortifications was very elaborate with the intention of making it as difficult as possible for an enemy attacker to breach the defences. Most of Malta's fortifications are sited around the Grand Harbour, with the fortified city of Valletta on one side and the three small towns of Vittoriosa, Cospicua and Senglea, collectively referred to as Cottonera, on the other. The latter are defended by a complex network of fortified lines and structures.

Up until the end of the last century, no public funds were dedicated to urban conservation projects and virtually all of the vast network of fortifications was decaying and falling into disrepair. The only notable exception to this was the conversion of St. James Cavalier into an arts centre, the Centre for Creativity. The massive cavalier is a defensive structure forming part of Valletta's landward fortifications. Two circular water tanks within it were dug out to create a small theatre in the round and a spacious atrium. The Centre for Creativity also houses extensive exhibition spaces, a cinema, a restaurant and a café. The structure itself is of great interest because its construction is a reflection of the building construction techniques at the time of the Knights.

In 2004, Malta became a member of the European Union. Significant funds were made available to Malta for a wide range of projects and initiatives including projects intended to enhance Malta's tourism competitivity. Substantial funds were invested in the urban conservation and more specifically in the restoration and reuse of forts and fortifications. For all the fortifications projects, the most intensive period of implementation was between 2006 and 2012, with virtually all projects being completed by 2015. The projects were made possible with funds from the European Regional Development Fund. These were allocated on the proviso that the running of the newly established uses will generate enough funds to cover the eventual running and maintenance costs. The following are some of the urban conservation projects for which a significant proportion of the capital costs were covered by European Regional Development Funds.

*Fort St. Elmo:* This is an extensive fortification system occupying a large area at the end of the Valletta peninsula. It is strategically located overlooking the entrances of the Grand Harbour and Marsamxett Harbour. It is arguably Malta's most important historic site because of its role in two important events in the history of Malta and of the Mediterranean, the Great Siege of 1565 and the island's defence in World War II (1940–1944). Within the fort and the bastions are many buildings, mostly small in size, many of which were used as barracks. Most of the buildings were restored and converted into use as a military history museum. The restoration project also provided for a heritage walk along the bastions as well as a series of attractive spaces within the fort and along the bastions (figure 9.1).

*Fort St. Angelo:* The fort is physically prominent at the head of the Birgu peninsula within the Grand Harbour. Extensive restoration works were carried out. It is now a major visitor attraction highlighting its history and its military roles through the ages. Fort St. Angelo offers spectacular panoramic views of the Grand Harbour and its surrounding fortified towns.

*Cittadella, Gozo:* The Cittadella is built on a hill overlooking the town of Rabat in Gozo. Because of its location on high ground, it is likely that the

**Figure 9.1 Fort St. Elmo, Valletta.** *Source:* Author's own.

area was inhabited since prehistoric times. The current shape and form of the Cittadella buildings and fortifications date back to the 16th century. Apart from extensive restoration works, the Cittadella project provided for a better presentation of the Cittadella and a greatly enhanced visitor experience. The Cittadella project also included the creation of a visitor centre, the rehabilitation of public spaces and the provision of improved access.

*The Fortifications Interpretation Centre:* This is housed in a 16th-century warehouse and forms part of the Valletta fortifications overlooking Marsamxett Harbour. The project involved extensive reconstruction works and internal and external restoration works. The interpretation centre is a means for raising awareness of the military architecture heritage of the Maltese Islands.

*Restoration and Rehabilitation of Fortifications:* Significant ERDF funds were invested in the restoration and rehabilitation of extensive stretches of fortifications of Valletta, Cospicua, Mdina and the Cittadella. This greatly improved their presentation and accessibility and enhanced Malta's historic walled towns.

## 9.4 WORLD HERITAGE SITES AND HISTORIC CITIES

World Heritage Sites (WHSs) are areas or sites of outstanding universal value recognised under the World Heritage Convention. Although the concept of a common heritage was not new, its importance significantly increased after UNESCO's adoption of the Convention in 1972 and the resultant global acknowledgement of such a concept. The Convention is a wide-ranging document that articulates its understanding of heritage and explains how properties that are included on the World Heritage List should be protected.[52] As of October 2020, there are 194 states across the globe that are signatory to the Convention. Since the first twelve inscriptions on the World Heritage List in 1978, the number of inscribed sites has gradually increased reaching 1,121 in 2020, including both cultural and natural sites.[53] The awareness of the 'common heritage of mankind', and the need to protect it, gained momentum after a global war that brought massive destruction in many cities and the subsequent radical remodelling of city centres and other urban areas brought about by industrialisation and urbanisation.[54]

It is national governments that nominate sites for inclusion in the World Heritage List. The final decision for inclusion, or non-inclusion, is taken by the intergovernmental World Heritage Committee (WHC) based on the advice of heritage experts. In the evaluation and decision process, a place that was recognised as locally and nationally significant is given an additional layer of meaning.[55] The concept behind the World Heritage inscription is that certain places on earth are of outstanding universal value and as such should form part of the common heritage of humankind. Some parts of the world's cultural heritage are so unique and important globally that their protection and conservation for present and future generations is a matter of concern not only for individual nations but also for the international community as a whole.

Inscription of a property on the World Heritage List brings with it responsibilities on the state; responsibilities that are monitored by the UNESCO's WHC. It does not give added legal protection to the property per se but draws attention to the importance of the particular site and places the responsibility on the national government to ensure its protection. The World Heritage List reinforces the identification, preservation and transmission towards future generations of places and monuments that are considered to be of outstanding universal value.[56] A main objective of the inclusion of cultural and natural heritage sites in the World Heritage List is to provide a mechanism for their protection and conservation.[57]

Several towns and cities across the globe are included in the UNESCO's World Heritage List, including many in Europe. The World Heritage concept unites diverse cities and urban areas who share the same opportunities and challenges associated with its recognition. These relate primarily to conservation and regeneration but there are also challenges related to the symbolic aspects and the meanings that are associated with sites. Appendix B includes information on urban areas and towns that are WHSs.[58] For each, a brief description is provided[59] together with the year of inscription, the area of property (in hectares) and the criteria for the inscription. Virtually all are urban areas with hundreds years of history. Some of the WHS are historic urban areas of less than thirty hectares, located in much larger towns or cities. Examples include the historic centre of Banská Štiavnica (Slovakia), the 'City of Luxembourg', 'Old Rauma' (Finland) and the historic centre of Urbino (Italy). Other sites included in appendix B are entire towns or cities, such as Venice and Valletta in Malta. In total, sixty WHSs are listed in appendix B, with virtually every country in Europe having at least one historic urban area that has been inscribed as a WHS.

The WHC presents and interprets the pre-eminent heritage sites of the world. The actual care of the listed monuments is left to individual states. Together with the International Council on Monuments and Sites (ICOMOS), UNESCO enacted conventions and charters to define and regulate governmental and professional management and conservation practices. These are endorsed by almost all national states globally. With this process, the WHC reaffirms the concept that heritage sites are cultural property belonging to all people. It enables local sites to become global.[60]

The actions of the WHC aim to create a new community based on heritage, mutual understanding and acceptance of cultural diversities. The World Heritage concept may be considered as a form of intercultural communication that targets acceptance and integration of others, as well as the creation of a space for cultural differences and for mutual understanding. This makes it an instrument for peace largely by seeking to depoliticise heritage. Such an approach though does not exclude the use of heritage by local groups seeking to maintain their identities[61] and this may result in tensions between neighbouring ethnic communities.

For a property to be listed as a WHS, the relevant authorities of a country (referred to as the 'state party' in WHC literature) submit a nomination. This is scrutinised by teams of experts who prepare a recommendation to the WHC, which then takes a decision. For a nomination to be successful, the property has to meet at least one of the criteria, preferably more, to demonstrate that it possesses outstanding universal value to humanity that merits protection for future generations. This has to be demonstrated in the nomination. Moreover, the nomination has to include a management plan that shows the significance of the site, identify the benefits of WHS inscription and demonstrate the proposed actions to protect the site. The WHC encourages nominations to involve the partnership of different stakeholders and in particular for the nomination to have community support.

The criteria for WHS inscription are listed in the Convention. There are six criteria that are relevant to cultural properties. For the full text of the criteria for inscription, refer to appendix A. In summary, these are as follows:

(i) a masterpiece of human creative genius
(ii) an interchange of human values, on developments in architecture, monumental arts, town-planning or landscape design
(iii) a unique testimony to a cultural tradition or to a civilisation
(iv) an outstanding example of building ensemble illustrating a significant stage in human history
(v) an outstanding example of a human settlement representative of a culture
(vi) an association with ideas or beliefs; or with outstanding artistic and literary works

In the case of cities, towns and urban areas, the criterion that is most often used to justify the outstanding universal value is criterion (iv) which says 'to be an outstanding example of a type of building, architectural or technological ensemble or landscape which illustrates (a) significant stage(s) in human history'. Also frequently used is criterion (ii), that is, 'to exhibit an important interchange of human values, over a span of time or within a cultural area of the world, on developments in architecture or technology, monumental arts, town-planning or landscape design'.

Having a WHS within a city brings to it several benefits. It makes the city and the country more prestigious and raises its international profile. An enhanced public profile is a means for attracting tourism and also inward investment. It generates local social capital in the form of increased public awareness of heritage and greater civic pride. For the property itself, WHS inscription could be a means for generating funds for conservation or for heritage-inspired regeneration. Over and above national legislation, inscription gives added protection to a site through the moral pressure that is exerted on state authorities to adequately maintain and protect it.

Inscription on the World Heritage List gives added reason for tourists to visit and hence, upon inscription, many World Heritage cities and sites experience significant increases in the number of visitors. Whereas a main objective of inscription is protection and preservation, it is paradoxical that some city destinations are prioritising their tourist development and using their World Heritage status for this purpose. In some cases, the increased tourism can put in danger the very survival of the World Heritage status,[62] in part because of increased pressures for tourism-related developments. These past twenty years, the WHC had to deal with an increasing number of situations where large-scale development projects were taking place within the boundaries of urban WHSs or within their buffer zones. These cities include Vienna, Edinburgh, Vilnius, Riga, Graz, Liverpool and Dresden.

## Case Study UC2: Suomenlinna, Helsinki: Conservation and adaptive reuse

Suomenlinna Sea Fortress[63] is located on a group of small islands off the coast of Helsinki, Finland. Suomenlinna was established as an island fortress first by the Swedish and then by the Russian occupiers. After Finnish independence in 1917, it continued to be used for military purposes. The bastion fortress was mainly built in the 18th century with improvements being made in the 19th century. In 1919, the islands and their fortifications were designated as a national monument under the Antiquities Act. This provided the impetus for restoration works, even if the islands remained under military control. In the early 1970s, its military use was substantially reduced and the islands were passed on to the civil authorities. The Governing Body of Suomenlinna (GBS) was set up with representatives of different ministries and agencies. The Board also includes two representatives elected by the residents. The GBS employs 70 full-time employees as well as additional staff during the summer. It runs on an annual budget of 12 million euros financed from property rentals and a state grant. In 1974, a master plan was drawn up, with the emphasis being on creating a resident community. Tourism and recreation also featured strongly in the master plan.

In 1991, Suomenlinna was inscribed in the UNESCO World Heritage List as an example of the 18th-century military architecture. The monument consists of two types of structures: the buildings and the lines of fortifications. Wherever the layout and form of the internal spaces allowed, the buildings were converted to residential. Where this was not possible, buildings were converted to other uses such as offices and conference facilities. The viable use of a building is the most effective way to preserve the monument in that the use will generate funds for its continued maintenance and conservation. With lines of fortifications, it is a different matter. Although fortifications may have some internal spaces, these are normally very limited and/or inappropriate for any modern day use. This makes it more difficult to identify a use that will generate funds for maintenance and preservation.

Apart from being a World Heritage Site, Suomenlinna is also home to about 800 permanent residents. When the islands were transferred from military to civil administration in the early 1970s, it was decided that the fortress would be maintained both as a museum and as a living part of the city and therefore increasing the permanent population was considered important. Suomenlinna is generally considered an attractive place to live. It is often described as 'a small village that happens to be only a 15-minute ferry ride from the centre of the capital'.

Since the 1970s, there was an ongoing programme of projects to convert historic buildings to residential use and subsequently to renovate them periodically.

Many of the buildings were originally built as barracks or as housing for officers. For each conversion to residential use, unique solutions were needed in terms of layout and choice of material to ensure that modern day requirements were met, without compromising the historic value of the building. It was sometimes problematic to reconcile the provision of essential services (heating, ventilation, running water and waste water drains) with the safeguarding of the authenticity of the historic structure. Residents are provided with basic municipal services, including transport connection, the school and the day care centre. Other services available on the main island include a church, food shop, a sports hall, a library, a public beach and a sport field. Health services are only available on the mainland although a service tunnel allows for the passage of emergency vehicles. The creation of a resident community necessitates the creation of jobs. The islands provide 400 full time jobs and a further 100 seasonal summer workers.

Suomenlinna is one of the most popular tourist destinations in Helsinki. It has been receiving visitors since the opening of the first museum, Ehrensvard Museum, set up in 1930. It grew progressively over the years and received a boost with the opening of new services to coincide with the 1952 Helsinki Olympics namely a new restaurant and a new ferry connection. In 1963, a tourist landing fee was abolished greatly expanding the recreational use of Suomenlinna. In 1998, Suomenlinna Visitor Centre was opened providing a range of tourism facilities, including the Suomenlinna Museum, a tourist info desk, multimedia presentations, shop and a café. Suomenlinna Museum opens throughout the year whereas another five museums across the islands open during the summer months. The islands also include facilities for conferences and functions. This allows for the viable use of internal spaces which would not be otherwise amenable to residential conversion. In 2016, over 1 million people visited Suomenlinna, mostly for short visits of 2 to 5 hours. This represents a 50 per cent increase in 10 years. The share of international visitors to Suomenlinna increased from 17 per cent in 1997 to 57 per cent in 2014. There are two major visitor groups: residents of Helsinki who spend leisure time walking there and tourists who come to see the fortress.

Inevitably high numbers of visitors in a relatively small area raised concerns about sustainability. The GBS drew up a sustainable tourism strategy for Suomenlinna in 2006, and then again in 2015. This was prepared in consultation with local residents, travel and tourism businesses, Helsinki city government and transportation representatives. The underlying objective is to minimise the negative impact while taking initiatives to maximize benefits. One initiative was to create a visitor route connecting the more important attractions and services, and thus implicitly discouraging people from parts of the island that are more sensitive. Most visits are in the summer months from May to September. Efforts to reduce seasonality include the organisation of events on Suomenlinna during the winter. The GBS regularly communicates and consults with local residents and with the travel and tourism businesses as this is considered central to the sustainable tourism strategy.

The experience of Suomenlinna provides interesting lessons on how to reconcile conservation objectives with the social and economic needs of the community. Over a period of four decades, various conservation projects were carried out on different buildings and structures across the island. These projects were carried out within a well-defined planning framework set out by a master plan and other policy documents. The master plan, prepared and approved in the early 1970s, provides a sense of direction and ensures that each project is compatible with and supports the objectives as set out in the master plan. Inevitably a master plan would need to be updated and new policy documents are required like for example the 'A Sustainable Tourism Strategy for Suomenlinna' of 2015. Even if there are changes to the planning framework, the sense of direction for Suomenlinna's management and conservation remains more or less consistent. An essential element of this approach is that the master plan and all the projects on Suomenlinna are the responsibility of a single agency, the Governing Body, so that coordination and the reconciliation of competing demands is achieved within one agency, rather than between different agencies. The GBS is staffed with different expertise (historians, conservation architects, tourism practitioners and others) to ensure that it can adequately handle competing demands. The effectiveness of the Suomenlinna's planning framework is that it does not look at the historic buildings and structures in isolation. They are part of the landscape of the island that also includes various human activities as well as stretches of natural landscape and coastline.

*Chapter 10*

# Architecture and Tourism

## 10.1 THE ROLE OF ARCHITECTURE IN CITY TOURISM

Architecture is part of everybody's lives even if people are not aware of it. Many aspects of people's lives take place in a building that has been designed and constructed with a specific purpose in mind. The most common uses of city buildings are as residences or offices as well as for retail, transport, industry and worship purposes but there are also tourism-related buildings such as hotels, catering establishments and visitor attractions. The word 'architecture' is normally associated with buildings that are monumental or that stand out because of their design or their size. More recent commentators have adopted a broader definition of architecture to include everything that is built,[1] including the more mundane buildings in the city.

Attractiveness in a city is created by a mixture of buildings from different eras with different functions and styles.[2] Many cities provide diversified architectural structure from different eras of their history including a broad range of contemporary architecture. Examples include Berlin, London, Paris, Barcelona, Moscow and many others. There are however some cities that are bound to a few specific eras from their past. For cities such as Rome, Florence, Venice and Jerusalem, the dominant urban heritage and their tourist image are closely associated with a specific period of their history. They also provide contemporary architecture but these are mainly located at the outskirts of the city and are barely present in the tourist's mind.

Buildings are built for a specific purpose or mix of uses. But beyond that, architecture communicates meaning. In chapter 8, we discuss how buildings acquire meaning and explain how the communication of architecture is dependent on a process of codification whereby people understand messages on the basis of pre-determined meanings.[3] In tourism, semiotics is mostly linked with clear recognisable images. A tourist visiting the Alps will take pictures of mountains, timber cottages and cattle. Visiting San Gimigniano in Italy, a tourist will take a picture in which one or more of the towers typical of the place are included. A visitor to Paris will take at least one photo which includes the Eiffel Tower. This symbolises Paris and, in a larger context, French culture. According to Specht,[4] 'For the masses, architecture is a major element in the semiotics of tourism leading to the connotation of place'.

Contemporary architecture is unlikely to have a historical meaning or significance. Yet, it would be wrong to refer to contemporary architecture as meaningless. It is not uncommon for recent works of architecture to have meaning that is either intended when being developed or one that is acquired over time. The meaning of architecture does not necessarily evolve from historical events. Most architecture has been contemporary at some point in history while in time it may have gained, changed and sometimes even lost significance.[5] Meaning in architecture depends on many factors but mostly on the perspective of an individual or of the community. For example, for residents of Bilbao, the meaning of the city's Guggenheim Museum locals might range from cultural invader to economic redeemer to transformational activator. On the other hand, for visitors, the museum might stand for extraordinary contemporary architecture while the connection to the city of Bilbao itself plays a secondary role.[6]

Tourism cannot happen without architecture, more so in cities. It creates the basic conditions for tourism to happen as the facilities and services are provided in buildings or in urban spaces. Architecture plays a critical role in almost every area of tourism. It provides infrastructure to enable tourists to reach their destination and, upon arrival, accommodation where to stay. It also offers venues for leisure activities. Beyond the functional use of building and spaces, there are instances where architecture becomes a place that draws visitors by providing something of interest and hence becomes an attraction. Moreover, architecture can be a major motivator in the tourist's destination choice. Some historical monuments have been attractions since the early days of tourism. Some examples include St. Peter's Basilica in Rome and the Pyramids in Cairo, Egypt. Cities are attractive to visitors not only because of their iconic architecture designed by international renowned architects but also because of 'their overall design, harmonious composition of open spaces and built form, and streets with views and interesting or surprising features'.[7]

There are instances where a building is developed to address local need without any direct function related to tourism. In time, however, the building becomes a place of interest for tourists. There are diverse circumstances why this could come about, that can be broadly grouped into two. This could relate to physical features such as design, scale or architectural treatment (e.g. Casa Batlló in Barcelona). Alternatively, it could be because of a historical event with which the building becomes associated (e.g. the General Post Office in Dublin, which still bears bullet marks from the 1916 rebellion).

Some buildings of the recent past are considered as attractors in cities sometimes even more than the historic architecture. In most cases, however, the visit to the recent architecture may be a brief one in a city that offers many other attractors. The exception to this is Bilbao where a piece of modern architecture has become a strong motivator for people to visit the city. No building has increased the awareness of contemporary architecture as a tourist attraction as much as the Guggenheim Bilbao.[8]

The shift to a globalised environment has meant that cities are forced to compete with each other in order to be attractive tourist destinations, desirable workplaces and more. The awareness and image of destinations are amongst the most important factors regarding their competitiveness. Destinations without a perceptible face and a clear image often have difficulty to compete globally. Accordingly, visually distinctive attractions can provide a competitive edge if linked to a positive destination image.[9] Urban destinations need to reinvent themselves over and over again in order to remain attractive and interesting places for tourism.[10] The development of contemporary architecture is one possible means that cities can do this. Over the past twenty years, tourist cities such as Amsterdam, Barcelona, Berlin and Paris have invested heavily in contemporary architecture to further enhance their image and elevate their position in the perception of the world, attracting interest and investment far beyond the field of tourism.[11] Uniquely designed architecture captures and enhances the special local characteristics of place to which tourists are attracted. City authorities and cultural organisations are increasingly aware that architecture has the potential to be a visitor attraction in its own right.[12]

Skyscrapers are a form of architecture that merits some discussion in tourism academic literature. Increasingly, the skylines of more and more cities are being dominated with tall buildings and skyscrapers as they are physically prominent features in many cities. Politicians, corporate tenants and architects often believe that 'large, upwardly thrusting symbols have promotional and competitive benefits'.[13]

Skyscrapers are normally ineffective as tourism attractions. London provides some interesting exceptions to this. In a discussion on vertical urban tourism in London, Andrew Smith[14] considers several developments that provide opportunities to visitors to experience the city from up high. He lists a number of buildings and structures that allow visitors to consume London including viewing platforms in new skyscrapers or in regenerated industrial structures (e.g. SkyGarden, the View from the Shard and Tate Modern). There are also moving attractions that simulate flight, most notably the London Eye. Other attractions facilitate a more physical experience, such as climbing onto the roof of the O2 Arena or descending down The Slide at The ArcelorMittal Orbit in London's Olympic Park. In the past two decades, developments in London have capitalised on the aesthetic appeal of urban panoramas and the popularity of viewing platforms with tourists. The provision of tourism facilities in tall buildings and structures generates rent revenues and thus makes the development investment more financially viable. Panorama attractions enhance the public perception of the tall building and thus also enhance the rental value of the commercial spaces.

## 10.2 MUSEUM ARCHITECTURE

Cities have been increasingly looking for new ways to promote themselves. In efforts to distinguish themselves from competitors, cities often seek to present themselves as young, modern, contemporary, forward-looking, creative and stimulating and to do so they resort to a good 'cathedral of contemporary art'.[15] Such 'cathedrals' are normally museums located within easy reach of 'cathedrals of consumption' – shopping malls or the streets of commercial districts. Museums have been increasingly identified as catalysts for city attractiveness and also as means for giving a new image to the urban environment. This has led to a considerable increase in the number of museums and exhibition spaces of modern and contemporary art, often as part of urban regeneration projects.[16]

In an interesting book chapter on urban attractiveness, Guerisoli[17] discusses the architectural form of 'hypermuseum' describing it as 'a place where the container has become the content, like a work of art that attracts mass consumerism and use of space'. The museum acquires a monumental presence in the urban landscape and becomes a symbolic building linked to the city's urban identity. Hypermuseums are often hybrids that combine exhibition activities with other activities such as shopping, consumption of food and culture-oriented meetings and events. When a city commissions a hypermuseum, the intent, or at least the hope, is that the innovative architectural addition to the city will become a global icon. Examples of hypermuseums include the MAXXI: National Museum of XXI Century Arts[18] in Rome, Centre Georges Pompidou[19] in Paris, the MACBA in Raval, Barcelona, and the Acropolis Museum[20] in Athens.

Museum revenues from tickets, bookshops, cafes and restaurants have minimal effect on a city's budget. The presence of a new art centre however may have a powerful effect on the urban context through its architecture and its symbolic value. It is the outside, rather than the inside, where considerations about image and financial impact become important. A museum exceeds its cultural value if and when it takes on a flagship status.

Apart from its normal museum culture functions, a flagship museum attracts local, national and international visitors as places of culture, urban centres for leisure and city icons.[21] Their strong visual impact and their formal bold experimentation make them the emblems of the presence of contemporary arts in the city.

Two books[22,23] in architectural literature on museums describe new major museums that were built and came into operation in Europe and elsewhere in the years 1992 to 1999 and 1998 to 2004, respectively. Virtually all were designed by world-renowned architects and most have some quality or feature that potentially makes them iconic. In just twelve years (1992 to 2004), as many as twenty-eight new, potentially iconic, museums were developed across Europe as described in these two books. Given the major investment involved for each this is a remarkable number, reflecting the importance that cities are giving to this form of cultural development. In addition, there were museums and art galleries built before and after this period and others developed across Europe that did not attract attention in the architectural literature.

In the introduction to his book, Barreneche[24] captures the spirit of museum's role in modern society as follows:

> The beginning of this new century will surely go down in history as a golden age for the museum. Since an ongoing boom in museum building started in the 1990s, the public has greeted openings of new museums and expansions of older institutions with a previously unimaginable level of fanfare and excitement. Travelling exhibitions have become certifiable blockbusters as popular as hit movies and musicals. . . . The way museums now market themselves is a big reason for their surge in popularity. No longer elitist institutions, they vie with theme parks and other mass-market entertainment for a slice of the public's leisure time and disposal income. But an even bigger motive is, quite simply, the architecture.

In the past fifty years, two major museum developments brought about a radically new approach in the design of museums and a dramatic transformation in the museum as an institution. They also brought about a change in how museums are used as part of urban policy and are perceived by the wider public. Centre Georges Pompidou and the Guggenheim Bilbao were instrumental to bringing about this 'golden age for the museum.' Aspects of museum design of these two museums are discussed in Case Studies AT2 and AT3, respectively.

In an essay on museums, Stanslaus van Moos[25] reviewed major museums, new and older ones, and identified four different forms or architectural approaches to layout and design.

*1. The museum as a converted monument.* Many of the better-known, well-established museums were conversions of buildings that previously served as royal or ducal palaces. With one important exception, conversions from palace to museum did not attract attention as an architectural project. The Louvre Museum is the exception when its remodelling and modernisation, completed in 1989, generated significant controversy because of the introduction of a glass pyramid in the centre of the courtyard over the museum's spacious and welcoming new entrance.[26]

Other major museums and art galleries found their new home in former railway stations or industrial buildings; normally monumental buildings with a prominent location in the city. Examples include the Museum for Contemporary Art in Berlin. The former railway station was renovated and converted into an arts museum in 1996. The most renowned example of this type of conversion is the Musée d'Orsay in Paris housed in the former Gare d'Orsay. The railway station was built in 1900 and converted into an arts museum in 1986. With 3.6 million visitors in 2019, it is the second most visited museum in Paris after the Louvre. Other examples include the Deichtorhallen in Hamburg (formerly a railway station), the Tate Modern in London (housed in the former Bankside Power Station) and the Hallen für Neue Kunst in Schaffhausen, Switzerland (located in a former textile factory).

*2. The 'open' museum.* In its simplest form, museum architecture consists of 'four walls, light from above, two doors, one for those who enter and the other for those who exit'.[27] This thinking gives rise to a second museum form which is minimalist and which relinquishes all pretensions to architecture in the traditional sense. Mostly popular in the 1960s, it is hailed as 'open', democratic and user friendly by its proponents as it diminishes the 'threshold fear' that grips visitors as they enter into other 'hallowed' museum spaces.[28] The Centre Pompidou, Paris, is a spectacular example of this approach as its architecture did away with ornamentation and monumentality in museum design.

*3. The museum with traditional enfilades.* In architecture, an enfilade is a suite of rooms formally aligned with each other such that the doors of connecting rooms are along a single axis. This was a common feature in grand European architecture and is a common arrangement in museums and art galleries. Twentieth-century museum design never really abandoned this traditional approach of layout and design, even if it contradicts the 'open' museum concept. Associated with this approach is the application of ornamentation on the facades of the building, normally in line with classical norms. This approach is illustrated in the Sainsbury Wing of the National Gallery in London opened in 1991. Interestingly, a previous architectural proposal for the National Gallery extension was infamously described by Prince Charles (30 May 1984) as being 'a monstrous carbuncle on the face of a much-loved and elegant friend'.[29] The comment reverberated for many years in the architectural and cultural circles in the UK and it reinforced the polarisation between traditional and contemporary architecture styles and thinking.

4. *The museum as 'sculptural architecture'.* The museum interior consists of a series of organic and non-conventional internal spaces and forms that depart from the traditional museum design concepts. The external visual qualities and aesthetics of the building emerge from the juxtaposition of forms of its various elements. External ornamentation, for its own sake, is avoided. Examples include the Jewish Museum[30] in Berlin, the Phaeno Science Center[31] in Wolfsburg (Germany), Graz Kunsthaus[32] in Graz (Austria), the Imperial War Museum North[33] in Manchester (UK) and the Guggenheim Museum[34] in Bilbao (Spain).

## Case Study AT1: The redesign of the Military History Museum, Dresden

The Military History Museum (Militärhistorisches Museum der Bundeswehr) is located on the outskirts of the city of Dresden, Germany.[35] Following a radical redesign in 2001, it is now an important museum that seeks to present the causes and consequences of war and violence rather than glorify military armies and war. It was built in 1876 to be used as an armoury for the military. Within thirty years of completion, the building was converted into a military museum. With strong horizontal lines, the main neo-classical facade echoed the traditional and authoritarian thinking of nineteenth century architecture. The exhibits were chosen and displayed to reflect the idea of power and grandeur of war which the museum sought to uphold. Reflecting the region's shifting social and political positions over the last 135 years, the building served as a Nazi museum, a Soviet museum and an East German museum. The history and identity of the building was deeply rooted in war. It was closed in 1989 because of doubts about how the museum would fit into the history being created by German reunification.

The design process for the remodeling of the museum and building was initiated in the mid-1990s and completed in 2001. Architect Daniel Libeskind was entrusted with the design following a competitive process.

The redesign seeks to divert the focus of the museum away from military technology. While the then-existing collection was considered a valuable core around which to develop the concept, the new museum communicates to visitors the past and present impacts of the military on different parts of society. Apart from symbolic meanings, the redesign seeks to reconcile aesthetic aspects with functional needs including the comfort and convenience of visitors and the appropriate presentation and conservation of the artefacts. The museum drew two hundred and twenty thousand people in its first ten weeks in 2001.

Figure 10.1  Military History Museum, Dresden. *Source:* Rachel Ebejer.

The concept of the museum is first made apparent when approaching the building. A wedge cuts into the façade breaking the continuity of the façade horizontality. The wedge has a dramatic effect on the appearance of the building. The shape and modern material of the wedge contrasts sharply with the stone and the orderly design of the historic building façade. Within the wedge, a viewing platform enables visitors to look over the modern city. The wedge points to the area which the first bombs struck the city in the Second World War. The wedge symbolises two things. Firstly, it is an acknowledgement of Germany's history of war and violence. Secondly, it symbolises a break from the past and an end to the glorification of war. In every aspect the museum is designed to alter the public's perception of war (figure 10.1).

## 10.3 ICONIC ARCHITECTURE AND THE CITY

In many destinations, iconic architecture is part of the overall tourism product that is offered to tourists. What does icon mean? What is an iconic building? What makes it an icon? An icon can be defined as 'a person or thing regarded as a representative symbol or as worthy of veneration'.[36] An icon thus has an inherent positive and often emotional connotation. Landry[37] describes icons as 'projects or initiatives that are powerfully self-explanatory, jolt the imagination, surprise, challenge and raise expectations. In time they become easily recognizable'. For example, the Eiffel Tower is an icon as it reflects the confidence of Paris's role in the industrial age. Its size and method of construction offered a challenge and raised expectations for subsequent structures. To this day, the Eiffel Tower jolts people imagination.

Defining 'iconic building', rather than just 'icon', may be a bit more difficult. Sklair[38] describes it as one which is sufficiently innovative or famous to represent a movement, style or era. With reference to iconic architecture, Alaily-Mattar et al.[39] emphasise its two characteristics, namely its ability to generate identity to a place and the intention to capture attention. These, in turn, affect both the perceptions of visitors and the local population.

A limitation of Sklair's definition is that people's understanding of what is iconic is different depending on the own cultural and professional background. The extent to which a building is representative is subjective, just as meanings associated with buildings are subjective. Similarly, the extent to which a building is considered innovative is also subjective. What may be iconic to a person knowledgeable of architecture might be environmental degradation and 'just more of the same' to someone who is more focused on environmental issues. A further limitation of Sklair's definition is that it excludes innovative buildings that are representative of place.

In academic literature, there are various different terms that are used to refer to iconic buildings including 'signature buildings', 'destination icons' and 'cultural flagships'.[40] These are often used interchangeably. Other terms that are used interchangeably are 'flagship architecture', 'iconic architecture' and 'star architecture'. Alaily-Mattar et al.[41] draw a distinction between flagship and iconic architecture. The former refers to buildings and institutions that attract visitors with corresponding economic implications for the institution and the city.[42] The latter is an architectural genre.[43]

Architecture is very much a product of a particular time, place and people and therefore it is often perceived as a potent cultural and political symbol. Indeed, societies are often defined by their surviving architectural achievements.[44] Throughout history iconic buildings have reflected and displayed the aspirations and values of society. For example, in the middles ages, large and impressive cathedrals were built to emphasise the importance of religion to society. More recently, with the rise of consumerism and the demand for corporate identity, buildings are increasingly looked upon as images or marketing objects.[45] Buildings that acquire the iconic status are built to symbolise something greater than their intended functions. They also provide an architectonic 'fix' for the need to create meaning and value in societies in search of a new image and identity.[46] The emergence of contemporary iconic architecture is an expression of the long-standing desire of economic elites to materialise their power in urban space.[47] They do so by means of attention seeking buildings that are 'resources in struggles for meaning and, by implication, for power'.[48]

Cities competing for tourist dollars strive for distinctiveness in an increasingly globalised world, a distinctiveness that is frequently achieved through the making of new urban icons. In the global drive towards iconicity, architecture is no longer considered to be merely an object with a specific function but also in terms of its ability to bring about relevant transformations.[49] In an era of competition between cities architecture is being used increasingly as a tool for economic development, particularly through the development of spectacular and iconic buildings.[50] City authorities have increasingly sought to develop or emphasise new buildings to act as place symbols.[51]

In the past three decades, there have been constant increases in tourism activity across Europe. The architecture iconography is often used to develop images of place so, with expanding tourism, the construction of architectural icons has become an increasingly popular phenomenon.[52] Internationally recognised tourism icons are a powerful draw to any city destination fortunate enough to have inherited or created one. When using icons to attract visitors, the challenge is to make the icon uniquely associated with the city while at the same time making it internationally popular.

In many cases, the public rhetoric that accompanies the launching of attention seeking architectural projects suggests that the appointed star architect will design the best architectural solution and grant the needed visibility for the project and the city hosting it. The name of the star architect has become a key factor in the design of the project and more crucially in defining a positive image that can be communicated to a much wider international audience.[53] The star architect is encouraged 'to take risks, break the rules, upstage the competitors and shamelessly grab the spotlight'.[54] It is a unique combination of fame with symbolism and aesthetic qualities that creates the icon.[55]

In the past fifty years, two projects in particular set the tone for the development of other iconic architecture in Europe. These were the Centre Georges Pompidou in Paris and the Guggenheim Bilbao. These were already referred to in the previous section in the context of their impact on the design of museums globally (see also Case Studies AT2 and AT3).

It is not clear how an iconic building becomes significant for tourism nor how it eventually becomes part of the tourist imagery of the city destination. Being visually interesting and pleasing is one factor. This, in turn, makes it more likely for the building to be represented in the international media and this, in turn, encourages people to visit. Whether historical or contemporary, a radically different aesthetics attracts attention, even internationally. Another possibility is that at the time of their construction, the buildings looked like no other architecture. A similar outcome might even come about from chance or mistake, such as the Leaning Tower of Pisa in Italy.

Interestingly buildings that are iconic tend to be those that are popular and well known to tourists. It is a self-reinforcing cycle. Being visited by large numbers of people makes a building more well-known not only amongst visitors and potential tourists but also across a wider international public.

Behind the scenes, there are the promotional efforts of the owners of the building, the architects and of the city authorities, especially those responsible for tourism. The promotion adds fuel to the process as the building moves from anonymity to an iconic status, at least as hoped by the city authorities. The iconic building's promotion takes place in subtle ways and also more covertly. It takes place before, during and after construction. Controversies accelerate a building's path to iconicity as they get the building debated in the media and this makes it more widely known.

The more a building is visited the more it becomes iconic. And the more iconic, it is the more it is visited. This means that for a building that is sufficiently innovative and visually prominent, a museum or some other form of visitor attraction is more likely to be recognised as being iconic than, say, an office or residential building.

Another element that increases a building's potential for iconicity is the manner of engagement of the visitor with the building and its surroundings. There are many reasons why the Centre Pompidou in Paris became iconic, including the very unusual high-tech architectural design and the controversies that surrounded it. One element that made the building more interesting for visitors, and hence eventually more iconic, is the escalator on the front of the building. To enter the different levels of the building, people ride on the escalator which takes them from level zero right up to the top level. As the escalator is moving up, the visitor's perspective of the urban piazza below changes progressively from close street level view to an almost bird's eye view from the top. Another example is the Aarhus ARoS Art Museum in Denmark that has a 360-degree enclosed viewing gallery on the roof. Visitors spend time strolling around the walkway and taking in the views of the city below.

Marques and McIntosh[56] emphasise the role of context in iconic architecture as a means for developing an image and an identity for city. To create an image, the design of the architectural icon needs to consider context and character of a place. Iconic architecture is a more powerful tool for city branding if the building is site and context specific, and hence not reproducible.

In reality, there are few icons of contemporary architecture that have world recognition, even if many cities have made it their objective to create a globally recognised icon. There is also a growing concern that in a world of attention, deficit we are being subjected to 'icon overload'. The icons built in secondary cities normally acquire regional and possibly national recognition but not global attention. This is partly because the cities are not sufficiently known at an international level. Across Europe, there are many examples of highly innovative architecture, some of them by star architects, that could arguably become global icons but that remain relatively unknown, other than in architectural circles. Two examples are Temppeliaukio Church, Helsinki and the Military History Museum, Dresden, both of which are described in more detail in case studies in this chapter. Other examples include Havenhuis (Antwerp), Kunsthaus Graz, City Gate and Parliament Building (Valletta, Malta), The Deep (aquarium, Hull) and The Titanic Museum (Belfast). Iconic status accrues more easily to major developments in global cities such as Paris, Berlin and London. Secondary cities have to try much harder in a world where people are constantly being bombarded in cyberspace with information and images.[57]

Smith and Strand[58] argue for more modest and nationally oriented, rather than globally oriented, objectives. If cities try too hard to develop a global icon potential cultural, regeneration and experiential benefits may be compromised. For contemporary architecture, they argue in favour of a domestic focus and the provision of attractive public spaces. Modest initial submissions may, in time, result in a project that attracts international attention. Oslo's National Opera and Ballet House is a case in point (see Case Study AT4).

## Case Study AT2: Centre Georges Pompidou, Paris

Centre Georges Pompidou was opened in 1977 in the Beaubourg district of Paris. It was not referred to as a museum but 'as a centre open to forms of contemporary creation' and its function was said to be that of 'a meeting place, open to the city and its inhabitants'.[59] Half the site area was designated as a public square in front of the building. This was conceived as a place for socialization and in fact many people congregate in the square, especially during the opening times of the Centre. The architecture was a radical break from the past, especially from past museum architecture. In their design architects Renzo Piano and Richard Rogers rejected the idea of the building with the classical treatment of facades and adopted a high-tech industrial style. Building elements such as electrical, hydraulic and air-conditioning installations are normally located out of sight on the inside of the building. Piano and Rogers turned the building inside-out and put all structural and services facilities on the outside of the building making them clearly visible from the surrounding urban spaces. The service installations stand out in four colours according to their function: blue (air conditioning), yellow (electrical) green (hydraulic) and red (escalators and elevators). The placing of building elements on the outside freed up the interior to create large open spaces providing for flexibility in the way the interior is subdivided and used. Inevitably the radical design generated significant controversy with some critics arguing that the exceptional industrial style did not fit into the Beaubourg neighbourhood. The architecture was also criticised as it gives priority to form over function, thus compromising its functionality as a museum. Open plan layout and free standing temporary walls made it almost impossible to show sculpture and painting satisfactorily.[60] In fact after just eight years, a major overhaul was carried out. More conventional, solid galleries were created and some of the original high-tech features removed.

Of the many thousands of daily visitors, less than one in five actually enter the art museum section of the building with people preferring to just meet, hang-out and walk through. Some argue that, like many public art centres, the building serves mostly as 'a culturally legitimated amusement park and culture café'.[61] There are several counterarguments to this criticism. The building's role as a visitor attraction is an important one as it has helped in the regeneration of the Beaubourg district and generally added value to the overall tourist experience of Paris. Moreover, the way the public, including tourists, use the building is one step towards democratization of culture as the casual visitor may in time be encouraged to enter the museum proper. Finally, the ancillary spending in bookshops, souvenir stores, restaurants and franchises (which are not a prerequisite of exhibition entry) generates much-needed funds that enhance the financial viability of the cultural institution.

Another aspect of the Centre Pompidou is the way it has regenerated the Beaubourg neighbourhood – an area which until then was largely neglected by the authorities and which was considered as one of the least desirable areas of Paris. The development generated a significant influx of visitors into the area as well as investments in commercial and residential properties. These greatly enhanced the area's overall image. Centre Pompidou set a precedent in that, since its opening in 1977, museum developments in Europe are expected to act as catalytic agents of urban transformation and not just function as repositories of the arts.[62]

## Case Study AT3: Guggenheim Museum, Bilbao

The Guggenheim Museum in Bilbao, Spain, opened in 1997 as a cooperative venture between the Guggenheim Foundation and the Basque regional administration of northwestern Spain. Designed by internationally renowned architect Frank Gehry, the museum building consists of a series of interconnected structures whose extraordinary free-form mass suggests a gigantic work of abstract sculpture.[63] The interior space is organized around a large atrium. The curved forms required the use of computer-aided design using software programmes that were originally developed for the design of aircrafts. The use of computer software made it possible to design and shape every element including the titanium panels that provide the external skin to the building.[64] The museum building derives aesthetic energy from its location on the edge of the water, thus making it not only a piece of architecture but also an exercise in landscape design.[65]

Many tourists visit Bilbao mainly, or possibly only, to see the Guggenheim. It is by far the most well-known site in the city. Moreover, many visitors to Bilbao are attracted by the architecture of the building and not necessarily by its museum function. Some visitors opt to view the building from the outside only instead of visiting the museum and the exhibits.[66]

Substantial impacts on Bilbao's tourism economy can directly be traced back to the Guggenheim. It was estimated that in its first three years of operation, the museum helped to generate approximately 500 million dollars in economic activity.[67] In 2012, the number of visitors surpassed 1 million.[68] Of every five visitors, four are international tourists or visitors from outside the Basque region. This development is particularly interesting as it achieves a phenomenal success due to a combination of stunning architecture, a big name collection and huge amounts of publicity.[69] Bilbao successfully used the Guggenheim to regenerate itself; giving rise to what has been termed as the 'Guggenheim effect'[70] whereby entrepreneurial city authorities commission world-renowned architects to design eye-catching buildings with the explicit aim of positioning the city favourably in the global competitive environment. This remarkable piece of architecture epitomizes the new wave of iconic urban buildings which have been designed to reposition cities on the global stage and act as a focus for economic regeneration initiatives.[71]

The Guggenheim Bilbao was not just an audacious architectural achievement. It immediately became synonymous with an entire city and a symbol of regeneration of the city and the region. That a single building could capture so much of the popular imagination was a stunning architectural surprise of the end of the twentieth century.[72]

## Case Study AT4: Norwegian National Opera and Ballet House, Oslo

Oslo's new Opera House[73] is located in the waterside borough of Bjørvika, an urban area that overlooks the fjord. It was previously a run-down industrial area partly occupied by drug users and prostitutes. The project completed in 2008 is an example of how cultural projects are used for the regeneration of undesirable areas in cities. The overriding justification for the project was to provide a dedicated venue for cultural performances, opera in particular. But beyond the functional and artistic objectives, there were other objectives including to enhance the image of Oslo, to encourage further tourism activity and to act as a catalyst for the regeneration of the urban area of Bjørvika. As is the case with many iconic developments, planning for it was characterised by significant debate and controversy – mostly about location and cost. Many politicians and cultural operators preferred a central location for the Opera House as they were more interested in the cultural legacy rather than the project as a means for urban regeneration. In 1999, the Norwegian Parliament decided in favour of the Bjørvika site over two other alternatives. The project cost was another controversial aspect of the project. Public funds were needed not only for the capital cost of the project but also for subsidies to cover part of the operational cost. Eventually the cost of the building and the adjoining spaces was 400 million euros. It requires 50 million euros annually to operate. The cultural venue generates income from ticket sales, sponsorships, retail and catering leases and donations but the revenue generated covers only 25 per cent of running costs. The rest of the running costs are covered by public finances.

Interestingly those supporting the project for cultural reasons recognised that the urban regeneration objective provided added justification for the government to cover the significant costs. National objectives were also ascribed to the project, in particular to develop and reinforce the nation's identity and to promote citizen's self-esteem. This is also reflected in the official name given to the building – Norwegian National Opera and Ballet House. Research carried out on the project[74] shows that a recurring theme is ownership, with many interviewees arguing that the new cultural facility belongs to all the people of Norway and not just to the cultured elite. The Opera House is now a leading visitor attraction in Oslo, especially for residents of other Norwegian cities when visiting the capital. The Opera House is relevant to international tourism in that it may tip the scales in favour of Oslo when a potential tourist is considering which city to visit in Scandinavia.

More than just visual, the building and the area also have a wider experiential dimension. People go there not just to see the building but also to experience the space around it and over it. They go to walk, to sit, to relax and to look out towards the fjord. The urban setting that has been created fits with Norwegian priorities for good public space and 'turn rhetoric about reconnecting cities with their waterfronts into reality'.[75]

## Case Study AT5: The Temppeliaukio Church, Helsinki

Temppeliaukio Church[76,77] is located in the heart of Helsinki in a large rocky space surrounded by apartment buildings. The rock outcrop was kept as intact as possible by embedding the church into excavated rock thus preserving the openness of the setting and allowing public access onto the rocky ground above. The church is covered with a shallow copper-lined dome. The dome seems to float over the church interior, supported by narrow beams around the entire circumference connected to the surrounding the rugged rock wall. Inclined areas of roof glazing around the dome shed light onto the rock walls of the church highlighting the contrast between the irregularities of the rock walls and the exact geometry of the dome structure. In a similar fashion, the altar piece at one side of the church space is lit by natural daylight from above. With about half a million visitors annually, Temppeliaukio Church is a must-see sight in central Helsinki.

*Chapter 11*

# Cities and Events

## 11.1 WHAT ARE EVENTS?

Planned events have an important role in human society and have done so throughout history. Before the industrial revolution,[1] the daily agrarian activities were interspersed with festivals on special occasions. The tedious daily routine was broken up by events of all kinds. Countries across Europe have a rich tradition of rituals and ceremonies extending over thousands of years. The role of events in society and in cities is of considerable importance even if the historic driving force behind them has changed. Events oriented towards religions have become less important but carnivals, fairs and festivals are popular at various times of the year.[2]

Many of today's events are influenced by past traditions, even if these have been adapted over time in line with changes to society including urbanisation, industrialisation and an increasingly multicultural population. With developing immigration, particularly after World War II, settlers brought their own customs and tradition that have, in some cities, become part of the place's heritage. In Britain's case, the cultural collision with the first migrants from the former colonies of India, Pakistan and the Caribbean caused new traditions to be formed alongside the old.[3]

The modern urban festival had its beginnings in the nineteenth century with the Bayreuth Festival in 1876 in Germany, followed some decades later by the Salzburger Festspiele in 1920 in Austria.[4] Nineteenth-century cultural festivals were concerned with displaying what was regarded as high art by the upper classes of European society. Another type of event – the international exhibition – saw its beginnings in the nineteenth century. Perhaps the first such hallmark event was the Great Exhibition held at the Crystal Palace in London in 1851. The exhibition was staged to celebrate industrial achievements and to display a range of products from different parts of the world. The exposition lasted six months during which over 6 million visitors viewed the hundred thousand exhibits in a purpose-built exhibition space.[5]

In the post-war period, there was a shift in emphasis of urban festivals towards the promotion of unity and stability. This resulted in the establishment of several urban festivals, among them the Edinburgh International Festival and the Festival d'Avignon, which were both founded in 1947. Post-war European festivals generally still displayed a penchant for high art and a civilising aim, although some festivals experimented with the promotion of inclusivity and new forms of interaction between audience, artists and place.[6] Festivals and other one-off events were motivated, at least in part, by the need to boost morale. In 1951, the Festival of Britain was held along the South Bank in London in an area that had been badly damaged by the war. New facilities were developed, mostly culturally oriented, including the Royal Festival Hall. It was, in a sense, an early example of the waterfront cultural projects that would become a popular option for many post-industrial port cities.[7] In recent decades since the 1980s, there has been a process that can be characterised as 'festivalisation'. This is linked to the economic restructuring of cities and inter-city competitiveness. The urban festival was harnessed for more market-oriented purposes. The growth of the leisure and tourism industries has seen cities being increasingly marketed as objects of consumption with festivals constituting a central part of the image creation. Sometimes events are a means for 'selling' the city itself.[8]

Events vary tremendously in size and complexity: from the simple and small, such as the village feast, to the huge, complex and international. The larger the event the greater the complexity and also the more difficult it will be for the organisers to handle.[9] The uncertainty about cost, timing, technical requirements, attendance and potential impacts will be far greater in the organisation of an international major sports event than, say, the organisation of an academic conference or a wedding reception.

In a globalised world, mega events are of particular importance. These are events with a global orientation and which involve a competitive bidding process for a city to win the right to host it. Standing out as mega events are the Olympic Games and the FIFA World Cup, both of which are staged every four years in different cities. Major cities bid for mega events because they are a means for raising the profile of the city, promote tourism and create new infrastructure. Bidding success requires close collaboration between the national government and city authorities as well as other interested stakeholders such as major tourism operators. There are risks in bidding – not least the perception of waste of public resources if the bid is unsuccessful. The staging of mega sports events requires massive investment in infrastructure such as new stadia, revitalisation of public spaces and improved public transport.[10] Much of the funding is provided by the national government. The increased revenue that is generated to the private sector, primarily from tourism, is the economic justification of such significant public expenditure. Other sporting mega events include the Commonwealth Games, the Rugby World Cup, the Cricket World Cup, Formula 1 Grand Prix and various international football tournaments. Other than sports events, there are the World Expos and the European Capital of Culture that can also be considered as mega events.

A hallmark event is another type of large event. The difference from mega event is that the hallmark event evolved within a city or country and eventually the event and the city become inextricably linked and cannot be separated. There is often a mutually beneficial relationship between hallmark events and the host community. Getz[11] describes hallmark events as iconic in that they hold symbolic value. Smaller events can also be iconic when they hold special meanings for interest groups and subcultures. Examples of hallmark events include Malta's Middlesea Race, golf tournaments at St. Andrews in Scotland and the Wimbledon Tennis Tournament in the UK.

Broadly speaking, there are three distinct types of events that are staged at a city level or above; sports, business and cultural. There is a very broad range of sports events involving competition between teams or individuals. For most types of sports, there are national championships and also international competitions in which teams from different countries participate. Some sports such as football, basketball and tennis attract many spectators to the arena or stadium and hence also attract a lot of media attention.

In contemporary society, business events have become a means to sell a variety of products. Events facilitate information and economic exchange; from simple markets where things are bought and sold, to world fairs and all kinds of meetings and conferences. The term 'MICE' – meetings, incentives, conferences and exhibitions – is sometimes used to describe this cluster of event types. These are events that facilitate business and economic exchange.[12] Refer to chapter 6 for more on venues for conferences and exhibitions.

There is an important consideration to be made in relation to both sports and business events. Since such events bring people together their social aspect is important. Other than for the smallest events, all sports and business events include ancillary activities for participants, spectators and, where applicable, for the participants' families and friends. Sports and business events have a cultural element in that many of these ancillary activities are based on a particular feature of the city or a particular place (e.g. dinner offering typical gastronomy of the city or a walk tour of the historic city centre).

## 11.2 FESTIVALS AND CITY SPACES

The past three decades or so have seen a remarkable increase in the number of planned events and festivals in cities throughout Europe and elsewhere. It is impossible to determine accurately how many events are held across Europe as the number of annual festivals in many larger cities runs into hundreds. Many events are staged in purpose-built venues. Others make use of a city's public spaces.

The urban festival is a popular approach for creating experiences in urban spaces. It is an occasion in which people express their collective belonging to a group or a place. Festivals engender a community by creating opportunities for drawing on shared histories, shared cultural practices and ideals, as well as creating settings for social interactions.[13] Festivals offer opportunities for expressions of local knowledge, communal identity and shared values.[14]

The term festival refers to a series of events that are based on the idea of celebrating a specific theme, organised for a specific period of time and to which the public is invited to attend. This celebration can be held annually or less frequently and includes one-off events. Arts and cultural festivals, in particular, involve the celebration of themes relating to different cultural areas such as poetry, painting, film, dance and photography.

The festivalisation of a city is not just about an increase in the number of events being held in the city. It reflects a qualitative change intimately linked to the spatial and economic restructuring of the city. The mere staging of single events is insufficient for a city to move from being a city 'with events' to becoming an 'eventful city'. An integrated approach is needed between policies relating to the staging of events and other urban policies of the city with the specific objective of maximising the benefits of the event programme as a whole. Cities rely on events to sustain the experience economy. For many cities, events are no longer merely a peripheral diversion from the everyday business of the city. They have become their prime concern.[15]

Cities are natural locations for the staging of events. It is in a city's urban spaces where people can come together to share in an event; be it a park, a large square or a wide boulevard. Large numbers of people are in close proximity whether they are city residents in residential areas or tourists staying in hotels. Most can get to the event by walking or a short public transport journey. The city often serves as a backdrop to the urban festival with alternative spaces opened up and extra-ordinary performances staged in places usually reserved for more mundane activities. Urban space is more than just a place where the event experience occurs. It also helps shape and gives sense to that experience. Space adds value to experiences. An urban festival often uncovers hitherto neglected or hidden features and thus serves to show the city in a new light. The festival is a means for redefining the city.[16] The Notte Bianca held annually in Valletta is a good illustration of this. It first took place in 2006 and is now a well-established event in Malta's cultural calendar. Historic buildings, churches and museums remain open till late and streets come alive with recitals, opera, jazz, poetry readings, exhibitions, dance, walk tours, street theatre and more.[17] People get the opportunity to enter into historic buildings that are not normally open to the public except for a purpose related to their use. Listening to a classical concert in the courtyard of a historic palace or entering one of Valletta's many auberges[18] has, to some extent, redefined people's perception of the city.

As a city creates and stages more events it needs to make available more spaces in which to stage them. As events develop, they often expand to fill more space within the city than had originally been conceived. It may sometimes seem that an event creates its own space in the city, but very often it all comes down to public space management of urban space and the interactions between the event organisers and the city authorities.[19]

Many events take place in public spaces and therefore it is inevitable that their legitimacy in using that space is sometimes challenged. One example is the discussion surrounding the staging of the Notting Hill Carnival in London including issues relating to use of space, security and access. Changes are often made to the route to reduce congestion and reduce crime. The relevant authorities often debate about responsibility and payment for activities in the public space.[20]

When it comes to meaning, there is a reciprocity between the city and the festival, as the city imbues the festival with meaning, whereas the festival provides new meaning for the city it inhabits.[21] According to Richards and Palmer,[22] 'Events can be defined by the spaces in which they take place, but can in turn come to redefine those spaces'. Il Palio di Siena is a good example of the symbiotic relationship between an event and the urban space that hosts it. This is explained in Case Study EV1.

Cultural heritage is not limited to monuments and cultural artefacts. Though intangible, the events and activities that take place within urban spaces are part of the heritage of a city. According to the website of UNESCO's 'Intergovernmental Committee for the Safeguarding of Intangible Cultural Heritage', cultural heritage 'includes traditions or living expressions inherited from our ancestors and passed on to our descendants, such as oral traditions, performing arts, social practices, rituals, festive events, knowledge and practices concerning nature and the universe or the knowledge and skills to produce traditional crafts'. Intangible cultural heritage is fragile in that it can easily be changed or even lost. It is however an important factor in maintaining cultural diversity in the face of growing globalization. City festivals of minority ethnic groups promote an understanding of the intangible cultural heritage of different communities and thus promote intercultural dialogue and mutual respect for other ways of life. With a mechanism similar to that for World Heritage Sites, lists of Intangible Cultural Heritage are drawn up by UNESCO. As of 2019, there were 550 designations from 127 countries from across the globe. UNESCO's list of Intangible Cultural Heritage includes several festivals and similar events, some of which are in European cities.

One example is the Ommegang of Brussels. This is annual historical procession and popular festival takes place annually over two evenings in July in the historic centre of Brussels. It originated as a religious event in 1348 but then declined in the eighteenth century. It was revived from 1928 to 1930 based on historic descriptions. It consists of various ceremonies, a crossbow competition and a large procession that follows a one and a half kilometre route through the city's historic centre including the Grand-Place. The festival was inscribed on List of the Intangible Cultural Heritage in 2019.[23]

It is not uncommon for festivals and events to present aspects related to the traditions, practices, craftsmanship and art forms of the city and the region in which it is staged. The festival becomes a vehicle for the portrayal of intangible heritage to residents and visitors. For example, Flamenco is an artistic expression fusing song, dance and musicianship. In Andalusia, southern Spain, it is performed during religious festivals, rituals, church ceremonies and at private celebrations.[24] Flamenco also features in non-religious events and festivals in the Andalusian cities of Seville, Granada and Malaga in Spain.

### Case Study EV1: Il Palio di Siena

Il Palio di Siena is a good example of the symbiotic relationship between an event and the urban space that hosts it.[25] The Palio is a festival of medieval origin held twice a year in Siena. The main event is the bareback horse race that takes place in Piazza del Campo at the centre of the city. Lasting less than two minutes, the race consists of three turns around the piazza while thousands of spectators are packed into the centre of the piazza sharing in the thrill of the race. The jockeys dress in 15th-century costumes with the colours of the contrada, or city neighbourhood, which they represent. The rivalry and contestations between the contrade are intense.

Though the race is considered a secular event, each horse is blessed in the church of its contrada by the parish priest. A spectacular pageant, the Corteo Storico, is staged by representatives of the contrade prior to the horse race. The festival is enhanced by drummers and flag throwers who demonstrate their arts using the colourful banners of their respective contrada. The festival, including the race itself, are replete with ritual and tradition.

The square has a radiating paving pattern with the ground gently sloping from three sides down towards the fourth, thus making it possible for spectators to see most of the race from anywhere in the piazza. The piazza is surrounded by historic buildings of the twelfth and thirteenth centuries. The architecture is simple and harmonious. The form of the piazza and the buildings that surround it serve as a remarkable backdrop to the pageantry and ritual of the festival. The piazza is very much part of the imagery that people associate with the event. Consequently, the piazza has a acquired meaning and significance from the festival and the race.

## 11.3 WHY CITIES CHOOSE TO BE EVENTFUL

Both in private and in public, people feel the need to mark important happenings in their lives, to celebrate key moments. They are opportunities to socialise and celebrate achievements and for group identities to be formed.[26] In the modern world, socialising and business can occur virtually through various media. Notwithstanding this planned events retain their attractiveness as social occasions. People participate in events because of the opportunities for entertainment but also because of their sociability, festivity and the way they allow people to break from normal routines.[27] Events are excellent vehicles for experience reproduction because they are limited in time and imply co-presence, not just between producers and consumers but also the co-presence of fellow consumers. The shared experience of cultural events is often what makes them special which is one important reason why people attend to concerts rather than merely watch them on television. Beyond their individual and local community significance, events and festivals play a vital role at city and national levels.

There are many reasons why so many cities across Europe stage a large number of events and festivals. Cities use events as a means for achieving many public policy objectives. Many cities have realised that events have the ability to deliver a wide range of outcomes and enhance many different areas of urban life. Public policy objectives range from place marketing to urban renewal; from fostering social integration to encouraging healthy lifestyles.[28] Events have become a useful strategy for cities to reposition and differentiate themselves in an increasingly competitive world. A city in pursuit of inward investment will compete with other cities through urban entrepreneurial displays in the shape of festivals and events.[29] In choosing where to invest, potential investors consider a broad range of city characteristics, one of which is the quality of life – a factor that is dependent on the cultural offer of the city, amongst others.

A major cultural event can bring many benefits to a city's cultural sector. It provides for more creative activity and may also be the catalyst for the improvement of cultural community by means of added training, participation and experience. This will be achieved partly by increased interest from city residents in cultural activities resulting from the cultural event. In part, the increase audiences will be the result of increased number of tourists to the city. Audiences are essential for cultural venues and activities. In modern economies, it is increasingly difficult for public authorities to sustain culture through financing and therefore increased audiences, coupled with increased income through sponsorships, are essential.

Cultural events have emerged as a means of improving the image of cities, adding life to city streets and giving citizens renewed pride in the city. Economic benefits will derive from increased number of visitors to the city. The cultural event will generate increased spending by visitors and residents in cultural activities as well as in ancillary services such as catering, accommodation and transport. Increased economic activity also signifies more job creation in the cultural sector as well as in tourism. An interesting example is a medley of exhibitions that were organised in the Netherlands in 2019 to commemorate the 350th anniversary of Rembrandt's death. Exhibitions were featured at different times of the year in five different museums in three Dutch cities: Amsterdam, The Hague and Leiden, which is the city where Rembrandt was born.[30] Amsterdam's renowned Rijksmuseum was visited by 2.7 million people in 2019 – an increase of 17 per cent over 2018. Sixty per cent of visitors were international.[31] The increased number of tourists visiting Amsterdam during Rembrandt's commemoration year was reflected in increased visitorship to museums.

Generating tourism is often high in a city's list of objectives. In using events to attract tourists, cities can adopt one of two strategies. The first is to bid for a mega event in the hope that the bid is successful and that the subsequent staging of the event will enable the city to achieve its objectives. The second is to develop a range of small, medium and large events – including regional, national and international events – most of them on an annual basis. None of these series of events necessarily dominate but in combination, they project a city that is dynamic and exciting.[32] Cities that pursue the first approach and bid or host a major event strive to keep the momentum going by staging other smaller events – in line with the second strategy. Several writers argue that cities need to develop an events portfolio with a mix of large- and small-scale festivals and events. The portfolio should cover a variety of different disciplines, catering to different audiences and producing different outcomes.[33,34] Getz[35] points out that many cities had predominantly focused on single events because it can be a very complex and daunting task for a city authority to manage and coordinate dozens (or even hundreds) of events, to achieve a range of objectives and satisfy numerous stakeholders. Getz insists, however, that the future lies in portfolios of events, all managed together.

For many cities, the staging of events is an integral part of their place marketing strategy. The primary aim of place marketing is to construct an image of a place in order to make it attractive to current and potential residents, investors and visitors.[36] Both high culture and popular culture have become important sources for the images that are used to underpin the brand image of cities. In some cities, event images are now so important that they dominate the natural or physical features in the identification of the city. Owing to its spectacular and participatory character, the staging of a festival presents an appealing marketing device for policy-makers to enhance the city's image.[37] Animation, or 'vibrancy', is important to cities for a variety of reasons including economic ones. A lively atmosphere makes people feel good about living in a place and makes the city attractive to visit which, in turn, drives inward investment. Generating substantial tourist flows is a main objective for many cities that use festivals as part of their city marketing strategies.[38]

Major events are often promoted and marketed in mainstream media and through social media thus providing for longer-term tourism benefits. With mega events in particular, global media exposure provides the opportunity to plant and develop attractive destination attributes and images. A viewer watching, say, the Olympics or the World Cup might be influenced to visit the host city at a later date. The television coverage often includes short videos or 'postcards' of tourist attractions and experiences during the short breaks.[39] There might also be extensive destination marketing campaigns tied in with the event. Apart from the increasing number of visitors, the promotion will also raise the city's profile with an international audience and also with its own residents. The latter in particular will boost local pride and enhance a sense of belonging amongst city residents.

Using events for self-promotion and for attracting visitors is particularly popular with second-tier cities.[40] Such cities may have visitor attractions, including a pleasant historic core but they would lack the critical mass of pleasant places and visitor attractions that would be required to convince people to travel to them from afar. Accordingly, these cities look to events for their reinvention.

Beyond tourism and place marketing objectives in some instances, social and cultural objectives are more important. One example is the National Eisteddfod held every year in Wales during the first week of August. The Eisteddfod is a showcase for music, dance, performances, visual arts and literature – all being a reflection of Welsh language, arts and culture.[41] The presence of tourists is encouraged, not least to present Welsh culture to a wider audience. Tourism however is low in the event's list of objectives. The festival travels from place to place, alternating between north and south Wales, including the cities of Wrexham (2011) and Cardiff (2008 and 2018).

Another example is the Notting Hill Carnival that is staged in London to celebrate West Indian diaspora even if it is attended by many tourists.[42] Many city authorities hold the view that festivals have a role to deliver a wide range of social and economic outcomes. The process of festivalisation is not just about the delivery of arts and traditional culture. It has become a key element in the efforts of city and local authorities to promote community cohesion and greater acceptance of diversity.[43]

It is not uncommon for a city to develop a new major facility or venue as part of the preparations for a major sporting or cultural event. Urban regeneration of an area can be kick-started by developing one or more flagship venues in the area for the purpose of staging a major event. The event provides a good incentive to national government to help out financially in the development of new facilities, which would otherwise be unaffordable for the city. Events are increasingly being linked to the spatial development of the city. Events make a direct contribution to redevelopment through the upgrading and rehabilitation of spaces in the city.[44] An example of this is the Stade de France in Paris. The stadium was built as the centrepiece of the 1998 FIFA World Cup and was instrumental in the regeneration of St. Denis; an area just north of the centre of Paris.[45] St. Denis was badly affected by the loss of industry and obstructive transport infrastructure. The stadium site was previously a gasworks that had been derelict since 1980. Previous attempts at regenerating the area proved futile. It was the decision to use St. Denis for the 1998 World Cup that radically changed the area's fortunes as it brought state funding and leadership while also giving the private sector the confidence to invest in the area.[46]

An alternative approach that cities can adopt is the development of permanent high-quality physical landmarks.[47] Many medium-sized cities steer away from this option because of the lengthy development process and the high future operational costs. Instead they opt for events as a policy tool because it offers several advantages when compared to the development of physical landmarks. These include greater flexibility and greater ability to offer spectacles and atmosphere. Events are able to meet the visitor's need for co-presence and the feeling of 'being there'. Events can cost less and achieve greater impact in the short term.[48]

It is not just cities and destinations that use events to further their objectives. Individual attractions use events to increase visitation and broaden their market appeal. Many attractions use their venue and built facilities to create the right ambience for a wide range of themed events. Events become a regular part of their offerings to visitors.[49]

Newbold and Jordan[50] highlight a number of factors that are bringing about the localization of global festivals or, as they put it, their 'glocalisation'. They point out an increasing competition between cities and countries with more of such festivals being staged. With a greater choice, less people are likely to travel long distances for a festival and instead attend one which is within easy reach of home. This results in a greater proportion of festival attendees being from nearby cities in the region. Newbold and Jordan also note that there is a growing trend for festivals to rely on regionally based sponsors for funding. These are more likely to be interested in the economic and social welfare of local communities and the political and community support that this will bring. This makes regional sponsorship a major resource for global events. Local and regional political forces are more keenly aware of economic and other benefits that international events generate in their communities. For event organisers, cultivating good relations with the local authorities provide important guarantees: facilitation of the authorisation procedures, security, location and accessibility. Having good mutual relations increases the likelihood of indirect support in kind or through public subsidy. Global events take place within national frameworks which have some influence on a festival's configuration. The strength of partnerships with the cultural and social actors of a region is one of several factors that make a difference between countries. These various factors explain the dual global and local nature of world events.

Since 2000 public authorities are giving a much greater emphasis to the legacy of the event. Authorities are refocusing their events strategies towards long-term outcomes rather than short-term gain. Attention to legacy has made cities give greater attention to the post-event period. Rather than being just the inevitable outcome of events, legacy is now something that can be planned.[51] The tangible legacy of a major event refers to the provision of sporting and cultural facilities that continue to be used after the event in a manner that is beneficial to the city residents. There are however risks. Sporting mega events have a short time span but the stadia last for decades. Apart from the significant initial costs to build, large venues are expensive to run and to maintain. Post-event, it is difficult for a city to operate large facilities in a manner that makes them financially sustainable. The prospect of post-event under-utilisation of event facilities may result in accusations of wasteful spending – something city authorities would wish to avoid. It is widely acknowledged that new event venues should only be developed where there is proven need for them. Better alternatives are to upgrade existing facilities, set up temporary facilities or use the facilities of nearby towns. In spite of the financial risks, cities continue to develop superfluous new venues. Reasons include the need to impress the franchise holders, the desire to show off during the event and the naive assumption that once a venue is built it will generate sufficient interest and demand.[52] Greece spent approximately 9 billion euros on the Olympic Games in Athens in 2004. A decade after the event many of its venues now are under-utilised or unused. Post-event strategy plans are crucial for a long-term legacy. Public investments to host the Olympics may be justifiable to attract attention and to present the city's advantages in an inter-cities global competition. The legacy of such significant public investment is, however, a source of major concern.[53] The difficulty is to make the transition from the event to a regular ongoing profitable activity. This avoidance of the 'white elephant' scenario can be vital to financial viability and so every effort is made to find a suitable long-term tenant and use. This complicates the management of the stadium project as the needs of the event and the ongoing legacy have to be planned for simultaneously. One aspect of this issue relates to community acceptance. There is often popular support for a stadium if it is part of an Olympic or similar bid where city or national pride plays a part. When transferred to a private club, however, peoples' reactions often change to one of opposition. This was evident in the reaction to the Johan Cruyff Arena[54] when it changed from being part of an Olympic bid to the home of Ajax.[55]

### Case Study EV2: Galway Arts Festival

The arts festival is set in the western sea-facing city of Galway.[56] It was established in 1978 to celebrate and promote appreciation of the arts in the region. Initially inspired by social objectives the festival has expanded greatly from its modest beginnings to become one of the largest and most popular festivals in the Republic of Ireland. The event was responsible, in part, to bringing about developments in the city's cultural infrastructure which, in turn, facilitated a more commercial approach to arts production. The festival's growing popularity as a tourist attraction for both national and international audiences and its growing commercialism have created a series of dilemmas for the festival. Since the mid-1990s, it has been grappling with the tensions posed by trying to balance deep rooted, socially aligned artistic goals on the one hand with often conflicting economic imperatives on the other. The festival organisers remain committed to serving local needs but the city's changing social and economic contexts impose pressures on the way the festival is produced and consumed. The national media, the state tourism agency and major sponsors are major stakeholders of this event. They pressurise organisers to internationalise the festival programme as well as its audiences and this has made city residents concerned that the festival is losing its meaning for local people. Meanwhile the city's rapidly growing population and the increased number of tourists have created an unprecedented breadth and diversity of artistic needs. The growing stature of the festival as a tourist attraction, and also as a commercial enterprise, generates many benefits to the city, but at the same time, it is problematic for some local people.

## Case Study EV3: Stockholm Culture Festival

The Stockholm Culture Festival saw its beginnings in 2006.[57] The declared objectives of the city council for the festival were to promote the city's cultural heritage and to enhance its profile as a culturally vibrant capital. The long-term aim was to increase the number of visitors to the Swedish capital and enhance its cultural image. The first iteration of the festival was subject to political controversy which raised concerns whether it would be repeated in subsequent years but its success secured its future as an annual event. The number of visitors grew from two hundred and fifty thousand in the first year to three hundred thousand five years later and the number of events increased from four hundred to over five hundred.

The city had previously hosted the Stockholm Water Festival but it was discontinued in 1999 because it was seen as too commercial and rowdy. The culture festival was established in the vacuum that was left behind by the water festival with the explicit intention of avoiding the mistakes that had been committed previously.

The culture festival seeks to make the city come alive and create a space for residents and visitors to enjoy and experience the city through the cultural activities on offer. The festival includes an array of performances staged around the city over six days. They are mostly free to the public and aimed at a wider audience while also providing space for more avant-garde acts. The festival is organized around a number of public squares which constitute the main stages of the festival. Activities include children's events, films, visual art, performance art, walks and photography.

The intention is to encourage the participation of people including those who do not normally consume cultural activities. Culture is used as a vehicle for participation and social inclusion. The aim is to make culture widely accessible without compromising quality. For example, the Royal Opera offers outdoor taster performances during the festival. By bringing high culture out on the streets it seeks to show opera as not being elitist and that it is an art form accessible to all.

The main objective is to develop a positive reputation for the festival and for the city while, at the same time, developing strong links to Stockholm residents by exposing them to the rich cultural life of the city.

## 11.4 EUROPEAN CAPITAL OF CULTURE

The European Capital of Culture programme started in 1985. The idea was conceived by the then Greek minister of Culture Melina Mercouri who stressed the role culture could play in the development of Europe. The 1980s was a time when the concept of Europe was solely economic: embodied by the European Economic Community. The original intention of European Capital of Culture (henceforth referred to as ECoC) was to stimulate European awareness in order to support political unification and also to give Europe something akin to a 'soul'.[58]

The ECoC's declared objectives are to put cities at the heart of cultural life across Europe and to improve the quality of life in these cities through culture and art. Immler and Sakkers[59] list the benefits that a city can derive from nomination as ECoC. It generates extensive media interest that attracts increased tourism, boosts economic activity and stimulates greater interest in culture. Moreover, ECoC nomination fosters social cohesion and cultural participation and provides opportunities for the strengthening of the travel infrastructure and the regeneration of deprived districts. Through participation in year-long activities, citizens play a bigger role in their city's development and cultural expression and strengthen their sense of community. The ECoC seeks to raise the international profile of host cities and help promote and celebrate different European cultures. The ECoC is much more than putting up 'a year of culture' and the benefits of hosting the year reach further than those directly associated with culture.[60]

In a sense, the title of European Capital of Culture is a European 'quality label' for local culture. It is generally expected that a city's ECoC programme displays its own local and national traditions alongside shared elements of European culture. Richards and Wilson[61] claim that culture is considered as a glue that binds the EU member states together.

Interestingly, the ECoC programme did not start its life with clearly defined guidelines and selection criteria. Rather than imposing a prefigured model of urban cultural policy, it was a process that evolved over time with adaptations to address the needs and demands of the cities hosting it.[62]

A strong dimension of the project remains its appeal to candidates to be a 'European' city.[63] The ECoC event involved an initial period (1985–1989) where the focus was on the event itself and on the existing value of recognisable cities of great cultural import. The first city to be awarded the title was Athens (1985), followed by other important cultural cities such as Florence (1986), Amsterdam (1987), Berlin (1988) and Paris (1989). Initially, the emphasis was on traditional cultural exhibits with relatively small budgets and minimal planning effort. The format was of an exclusive art and culture festival. It was not considered necessary to set city-oriented objectives or to stimulate development as these cities were already capitals of culture by their very nature.[64] Since 1990, the rationale for staging the event shifted to one that prioritises the long-term effects on city-related matters such as cultural provision, levels of participation and urban development.[65]

Following Glasgow, there were other port cities in decline that sought to use culture to facilitate the regeneration of their ailing economies namely

Dublin (1991), Antwerp (1993), Rotterdam (2001), Genoa (2004) and Liverpool (2008). Although cultural factors were decisive for their selection, increased tourism was taken as the primary sign of success.

Apart from cities, wider urban areas were also selected as ECoC including Lille Metropole (2004), Greater Luxembourg (2007), Essen-Ruhr (2010) and Marseille Provence (2013). This gave the areas the opportunity to develop coherent strategies for entire regions, including several cities, rather than for just single cities. Moreover, many smaller cities on the fringes of Europe were awarded the title including Cork (2005), Patras (2006), Sibiu (2007), Stavanger (2008), Pécs (2010), Turku (2011), Guimarães (2012), Košice (2013) and Umeå (2014). This trend underlined Europe's diversity and emphasised that European culture is Europe-wide and not just restricted to the major cities. Up to 2019, there were sixty cities or city-regions across Europe that have held the title. The initiative has become one of the most prestigious and high-profile cultural events in Europe.

Bianchini et al.[66] note a linkage between mega events such as ECoC and urban regeneration. In a context of growing competition between cities, policy-makers often consider cultural mega events as part of a strategy to improve the image of a city, improve its infrastructure and revitalise its economy.

ECoC has developed into arguably the flagship cultural programme of the EU. This is because it requests an ever-growing number of candidate cities to address economic and urban regeneration on the one hand and social inclusion through civic participation on the other. The EU often experiences tensions in trying to pursue social goals while engaging in standard economic practices.[67]

The casual observer may consider the ECoC as consisting merely of a series of cultural activities that are held in a city in a particular year. The ECoC is much more than that as reflected by the reasons and objectives which motivate cities to bid for the title. Often ECoC is one of various urban policy tools used by cities to regenerate themselves. Urban renewal is achieved not only by implementing capital projects but also by breathing new life into a city's culture and by boosting tourism. ECoC is a means for raising the international profile of a city in a context of stiff competition between cities. The more dynamic cities implement strategies to strengthen their attractiveness and hence their competitiveness. Cities that are complacent risk losing out and become less renowned when compared to the more dynamic neighbouring cities. ECoC is also supported by the European Commission which sees it as a celebration of culture and a means for highlighting the richness and diversity of cultures across Europe.[68]

More recently there is a growing awareness of the value of exploiting ECoC as an opportunity to address citizenship and social issues. Cities bidding for the title still seem to reflect this ambition at the candidacy stage as can be witnessed by their submitted bid books.[69] Some cities put such ideals to the side in pursuit of more tangible and deliverable objectives like economic regeneration. This is particularly true in relation to the development of infrastructure, increasing growth and jobs and attracting higher numbers of tourists.[70]

For large-scale events such as ECoC, the long-term impacts are significantly more important than any short-term increase in tourism numbers. Long-term economic impacts can be achieved in one of two ways. First, ECoC has the potential to significantly alter the way the destination is perceived by a wider international public. For example, the cities of Glasgow, Genoa, Lille and Liverpool (ECoCs in 1990, 2004, 2004 and 2008, respectively) used the event imagery to visually break with the past and as part of narrative of 'rebirth'.[71] Second, ECoC is a vehicle for increased investment in infrastructure, be it cultural, tourism or transport. Lille 2004 (see Case Study EV5), for example, was characterised by a large-scale regeneration programme that included the development of new public spaces, the regeneration of historic buildings and new facilities for the staging of cultural activities.[72]

### Case Study EV4: Glasgow 1990, European Capital of Culture

Glasgow was designated as European Capital of Culture in 1990. This brought about significant changes to the way the title was given and how it would later be competed for.[73] Glasgow was arguably a challenging city well below par when compared to the cultural cities nominated in previous years. It was an industrial city in decline city with serious social problems and with a poor reputation in culture. Glasgow was nominated with the aim of redefining it through its culture, stimulating urban redevelopment, promoting the city's image and putting it on the European map.[74,75]

The event is widely perceived to be a success in changing the city's industrial profile and attributing a cultural image to the city.[76] Glasgow also demonstrated how an emphasis on culture can provide economic benefits. Glasgow's city centre experienced a considerable property boom with considerable investment particularly in cultural infrastructure in preparation for the 1990 ECoC. Modern art was included as a central theme for the event as a signifier of the shift away from the city's industrial past.[77]

Culture becomes a device that enables cities to be conceived as brands that attract visitors in search of experiences. Glasgow is credited with using European Capital of Culture 'to overhaul its image as a depressed, problem-ridden, post-industrial city into an attractive and culturally interesting service-driven contemporary city'.[78]

### Case Study EV5: Lille 2004, European Capital of Culture

Lille is arguably an unpretentious northern French city of two hundred thousand inhabitants located very close to the border with Belgium.[79] European Capital of Culture Lille 2004 included 193 towns in the entire metropolitan area. One of the objectives was to promote social cohesion and enhance pride and self-confidence of the residents. The emphasis of the programming was on having a large number of disseminated events targeting particular audiences and involving as many cultural producers as possible. Twelve new pole buildings for cultural activities, the Maisons Folie, were opened through suitable re-conversion of former industrial buildings. These were located across the entire metropolitan area and were intended to create a web of creative-friendly environments for artists and residents. The total ECoC budget was around 74 million euros with an additional 89 million euros for capital projects and historic building restoration. The investments projects included a restored opera House, two brand new cultural facilities and the restoration of several historic churches and monuments.

Lille was successful in mobilising people with substantial attendances to the events. Special tariffs were offered to residents and other classes of visitors. More than one thousand two hundred schools were involved in the programme, with nine hundred special events targeting very young audiences. Hundreds of small businesses hosted permanent infopoints, special openings, thematic furnishings of shop-windows and so on. Lille 2004 took a proactive approach to event legacy with the focus being on re-imaging this post-industrial city as a modern metropolis.[80] The number of enrolled volunteers in the 'ambassadors' programme was almost eighteen thousand including expatriates from all over the world. Several hundred people gave their direct contribution to at least one event with many being involved as full-time volunteers. 'Lille 3000: The Voyage Continues' came about in the wake of Lille 2004 following a decision by the organising team to extend the one-year programme into an ongoing event. It is a good example of how ECoC can generate a legacy with lasting benefits for the city. It was a deliberate attempt to leverage and capitalise on the cultural, social and economic impacts generated in 2004. Rather than being a regular event, it is referred to as 'sporadic cultural impulses' with four editions being staged in the 15 years following Lille ECoC.[81]

### Case Study EV6: Valletta 2018, European Capital of Culture

Malta[82] has an unusually rich urban heritage with historic fortified towns and numerous small historic villages cores.[83] Prior to the millennium, it was widely recognised that Valletta needed to be regenerated. There was a process of decline with more and more properties being left unused and allowed to decay. Public investment was limited to minor restoration projects. The lack of public investment was mirrored by a lack of private investment; creating a downward spiral and more dilapidated properties in many parts of Valletta. That changed in the first years of this century. Significant public sector projects were being carried out, especially in Valletta, including pedestrianisation and extensive restoration of fortifications and historic buildings.[84]

Since well before the millennium, the declared tourism policy was for the use of urban heritage to increase the share of culture tourists to Malta. In this context, European Capital of Culture Valletta 2018 (henceforth referred to as V18) had a particular significance. It provided the opportunity to enhance Malta's credentials as a culture destination. Preparations for V18 began soon after 2010. A lot of time and effort went into the preparation of the bid book[85] including extensive consultations with the cultural community and with the wider public. Cultural operators were encouraged to come forward with ideas for cultural initiatives.

The preparations for V18, together with major public investments, sent a clear message to the private sector that Valletta would in future be given the attention that it deserves. This was a catalyst for private investments in the restoration of Valletta's historic building, particularly for the conversion into boutique hotels and catering establishments.

The V18 programme was diverse, with as many as four hundred events. Several Maltese artists had the opportunity to collaborate and gain experience on large international projects. The down side was that the commitments made in the bid book were largely side-lined. The V18 management agency saw itself merely as a funding agency awarding funds in accordance to requests made. Changes were made to the V18 theme and to the cultural programme. These were most likely motivated by a desire to create more popular and crowd-pulling activities. Inexplicably major changes were made in personnel in 2014 and then again just a few months before ECoC year.

The situation was made more difficult when V18 agency chairman made some controversial comments on a highly sensitive subject.[86] He antagonised large sections of the Maltese public including the local cultural community. V18 was criticised by international stakeholders for words and actions that were not considered compliant to European values, especially that of inclusivity.

In the run-up, V18 was beneficial as it was one of several factors that encouraged private sector investment in the renovation of historic buildings in Valletta. The cultural programme was more varied and innovative when compared with previous years. It is not clear, however, to what extent these cultural benefits came from the staging of European Capital of Culture or whether the same benefits could have been equally achieved merely with an increase in financial allocations to Malta's cultural sector. Because of weak governance, the intended objective of enhancing Malta's cultural image did not materialise.

*Chapter 12*

# Post-Pandemic Prospects

## *Overtourism or 'Undertourism'?*

Prior to 2020, European tourism sustained steady growth for several decades. Global tourism, including tourism in Europe, has benefitted greatly from technological advances that led to cheaper airfares, easier means for people to plan and book their own travel, and the ease of sharing travel experiences with friends. The numbers of international tourists in Europe grew at 4 per cent per year or higher for seven consecutive years following the 2008/2009 global economic crisis. Spain, a strong European tourism performer, received 7 million more international tourist arrivals in 2016 to reach 75 million, a 10 per cent increase in just one year, following a decade of extraordinary tourism growth.[1] The overtourism that some cities experienced prior to 2020 has been extensively debated in tourism academic literature.[2,3,4] Overtourism incorporates a broad range of impacts including crowding at tourism hotspots and negative impacts on the liveability of residential areas. While acknowledging that that there were tourism hotspots that were significantly impacted, it is this author's view that the impacts of overtourism on cities were sometimes overstated and that many of the issues debated could be mitigated, or even resolved, with better urban management and more effective legislation. A 2018 study[5] surveyed eleven European cities to evaluate public perception of tourism in relation to a number of sustainability indicators including overtourism. Two of the cities surveyed, Barcelona and Amsterdam, identified overtourism as an issue and included specific urban policy measures in their tourism plans to address it.

Enter 2020 and with it the beginnings of a global pandemic. Suddenly and unexpectedly destinations that previously suffered from overtourism are now faced with the completely opposite concern: 'undertourism' or, rather, the absence of tourism.[6] The impact on the tourism and leisure industry during the pandemic has been devastating. Inevitably in the tourism industry there is a lot of debate on what the tourism situation will be post-pandemic and what the long-term impacts of the pandemic will be.

This chapter is being written in March 2021 at a time when many European countries are being hit hard by a third wave of the COVID-19 pandemic. It is a time when most countries across Europe are subject to significant restrictions or are in some form of lockdown. It is also a time when vaccination is picking up momentum across Europe. The crisis is far-reaching and it is very difficult to make predictions.[7] Notwithstanding this the subject merits a discussion here although it should be said that the situation will be much clearer by the time readers have the book in hand. In view of the complexity and uncertainty of the humanitarian and economic crisis, the best this author can offer is a few reflections on tourism and on wider societal implications. These reflections are based on various writings in the academic literature and on reports of international news agencies backed up with feedback from persons directly involved in the tourism industry.

Before debating the pandemic's long-term effects, it is important to highlight that this is not the first time that the tourism industry is struck by some form of crisis. Tourism of recent decades is replete with examples of natural disasters, economic downturns, political turmoil, health scares, terrorist activity and other events that have impacted negatively on the volume and direction of tourism flows. Tourists are typically averse to any kind of risk. Any actual or perceived threat to their health, safety or security is likely to influence their decision to visit a particular destination. A key determinant of a destination's success is its ability to provide a safe and secure environment for its visitors.[8] Even before the COVID-19 pandemic, there was a growing body of literature debating how tourist destinations should deal with a crisis situation in the industry. Such a discussion, however, is problematic because a one-size-fits-all approach to a future, undefined tourism crisis is of little use. Crises vary significantly in duration, scale and impacts.[9]

The COVID-19 pandemic has given rise to difficult issues of unprecedented complexity. In the first instance, this is a major health problem with authorities across the world struggling to keep down the numbers of hospitalisations and mortalities. Over and above, there are other health implications as non-COVID-19 patients are not getting the full medical services and treatments that they require. There is a higher incidence of mental health problems resulting from social isolation or financial stress caused by the pandemic. Superimposed on the health issues are the economic ones. Many households across Europe had a reduction in their income because of reduced work or even the loss of employment. Extensive public funds have been used to support families and businesses to help them get through the pandemic. By the summer of 2020, there had already been significant funds spent by governments to provide support. With a second, and possibly a third, pandemic wave, extending into 2021, the costs to national governments escalated further. The pandemic has affected all sections of society, in their health and finances, in every country and over an extended period of time. Hence the complexity and the difficult decisions that public authorities had to take since March 2020 and will continue to take in the coming months and possibly years.

It is very difficult to forecast the levels of tourism activity in 2022 and in the following years. When discussing the 2020–2021 pandemic situation, two words come to mind – complexity and uncertainty. The same two words apply to post-pandemic tourism. Uncertainty stems from so many questions for which, in the first quarter of 2021, there are no replies. Vaccination provides immunity from infection but for how long? Does vaccination prevent transmission? How much time will it take for European countries to be vaccinated sufficiently to achieve herd immunity? Is there the risk of some countries not achieving herd immunity if the vaccine take-up is insufficient? Will mutations of the virus undermine the effectiveness of the vaccine and jeopardize the attainment of herd immunity? The virus will not be eradicated completely and there will be pockets globally where the virus will keep spreading. With global travel can the coronavirus re-emerge in a European country months, or even years, after herd immunity is achieved?

Tourism and travel are closely interrelated. In particular, any discussion on the long-term prospects of international tourism necessitates a debate on the prospects of air travel. Will aviation recover to pre-pandemic levels? There are several factors that suggest that the full recovery of aviation, and hence tourism, will take many years. In 2021 and beyond, air travelers will be increasingly inconvenienced. COVID-19 tests, vaccination certificates, mask wearing, potential delays at airports and the risk of being stranded will all discourage foreign travel. Even after the major effects of the pandemic have subsided, elderly people and persons with underlying medical conditions will be wary of placing themselves in situations that they may consider risky. Many people's income has been very badly affected during the pandemic, leaving them with less money for non-essential expenditure such as holidaying abroad. Insurance costs will be higher to cover for potential travel disruption caused by COVID-related incidents. Corporate entities and businesses are likely to reduce their travel budgets and this will badly affect airlines that rely heavily on business travel to enhance their profits. Higher costs will dampen demand for travel and this, in turn, will further escalate the cost of air travel.

There are several factors that will mitigate against this downward trend in demand for air travel. First a significant pent-up demand for travel is being built up during the pandemic. For many people, travelling overseas is an essential element in their lifestyle. They will travel as soon as is practically possible, even if the cost may be higher than previously. Governments and city authorities are likely to be proactive and take initiatives to incentivise air travel. This is more so for cities and destinations where tourism is heavily dependent on air travel. Pre-pandemic, it was already common practice for public authorities to offer incentives to low-cost airlines to establish new airline routes to their airports.[10] Post-pandemic, and in conditions of low demand, the incentives offered to low-cost airlines are likely to be more extensive and attractive.

During the various pandemic lockdowns imposed across Europe, all tourism and related sectors came to a standstill. Even when and where there were no lockdowns, the restrictions were severe, forcing most businesses to operate at significantly reduced levels. Small businesses are the driving force in the tourism and leisure sector. Most small business were able to survive the first wave of the pandemic but as restrictions and lockdowns were reinstated for the second wave, financial survival became more problematic. It is difficult for small businesses to survive long periods of closure. Inevitably many small businesses are closing down or will close down in the coming months. The closure and loss of small businesses will hamper post-pandemic recovery in tourism destinations.

After the first wave of the pandemic in the second half of 2020 the events sector sought to continue operating by resorting to hybrid events. As the second wave of the pandemic struck globally, this was found to be impractical leaving virtual events as the only option.[11] Although attending virtually has some advantages it is generally problematic because the possibility to network and make contacts is significantly reduced. With virtual events, it is difficult for organisers to generate a profit. Moreover, people are beyond Zoom fatigue. Having the mental capacity to sit through a two-day long virtual event seems to be a stretch for most people. Most attendees to in-person events cherish the opportunity to take some time off from the event to visit places in the city where the event is taking place. It is fun 'being a tourist', even if it is for a limited amount of time.

In planning for post-pandemic tourism, public authorities should consider different tourism and leisure sectors as these are being affected differently. The following are the likely post-pandemic scenarios for different sectors. Clearly the uncertainty of the health and economic situations only allow very broad-brush projections.

In the events sector, conferences will continue to generate tourism to cities but in lower numbers as conference organisers resort to hybrid modes – combining the in-person event with online attendees. Major commercial exhibitions and events are likely to be affected in the short term as some companies and public agencies may decide to reduce their marketing budgets. Exhibition organisers will give greater attention to visitor management to avoid unnecessary crowding. These factors, combined with social distancing restrictions, will likely result in less visitors compared to pre-pandemic.

Music, theatre, dance and other forms of cultural performances will be subject to opposing pressures post-pandemic. There will be a high demand for such performances, not least to make up for what was missed out during the pandemic. There may be some people who will initially be wary of being part of a crowd in an audience but any fears will be quickly dispelled if there are no reports of infection in the city and region. Some public authorities may opt to incentivise the recovery of the sector to restore a city's social life and also to safeguard jobs in the sector. An expedited return to some form of normality in the cultural sector is vital to restore quality of life for city residents, even if culture-motivated tourism will take more time to recover.

In many cities, food and drink establishments are an essential part of people's lifestyle, primarily for purposes of socialising. Such establishments are likely to bounce back to pre-pandemic levels fairly quickly largely because of domestic demand. In cities that are dependent on international tourism, the recovery of food and drink establishments will be slower.

Except for during lockdowns, many hotels continued to operate during the pandemic even if at significantly reduced occupancies. Hotels have gained experience on how to make their establishment COVID-safe. The continued application of these measures will reassure clients. Once the pandemic is over, city hotels will recover fairly quickly although to levels that will probably be lower than previously.

Similarly, many museums and visitor attractions have introduced COVID safety measures and operated at significantly reduced levels when not in lockdown. The larger and better established museums will recover slowly, with COVID restrictions being eased gradually over time. It will take several years to reach pre-pandemic levels of visitations. For the smaller attractions, the situation will be more difficult as the added cost of COVID restrictions and the reduced visitorship will make it difficult for them to survive.

For most parts of central mainland Europe, a city has hundreds of other towns and cities, within travel distances of three or four hours using land-based transport such as rail or private car. Having visitors from the same country (or from adjoining countries, across the border) will enable cities to compensate for the reduction of international arrivals. The slowest to recover will likely be the cities and destinations that are heavily dependent on international tourism and which do not have a geographical hinterland that they can tap into for domestic tourism. Such is the case for cities at the periphery of Europe, especially those along the Mediterranean coast or on Mediterranean islands.

It is likely that the city destinations will take measures post-pandemic to reduce tourist crowding so that visitors will be able to enjoy the city in a more COVID-safe environment. Measures could include better management of visitors at tourist hotspots including setting limits on the number of visitors that can be in or near a tourist attraction at any one time. Another measure could be to divert tourism flows to other areas of the city by creating visitor attractions and enhancing the urban spaces.

An interesting initiative is being taken by the Uffizi Gallery in Florence to reduce overcrowding. At its busiest, the museum has 12,000 visitors per day resulting in crowding not only in the museum itself but very often in nearby streets and piazzas. The project is referred to as Uffizi Diffusi.[12] The idea is borrowed from the 'albergo diffuso' or the 'diffuse hotel' concept, in which individual tourism accommodation rooms are located in different houses of a village.[13] Artworks stored in the Uffizi's deposit will be put on show throughout the surrounding area of Tuscany. Towns and villages from across the region nominate buildings that could potentially become exhibition spaces. If successful, the initiative will help ease the pressures on Florence and address overtourism. Spreading out the supply of tourism resources across the city and the region will reduce crowding in the tourism hotspots at the city centre and thus make it and the Uffizi Gallery more COVID-safe. It will also enable tourists to 'discover' the charm of nearby towns – charm which would otherwise remain largely unnoticed by the international tourist. The first phase of the project should get underway in the summer of 2021.

Recovery will be faster for those destinations that are successful in implementing measures to make their city COVID-safe. Moreover, it is essential for destinations to communicate to their potential customers that the city, the hotels and the visitor attractions are COVID-safe.

The industry has undergone major disruptions in the past due to health, safety, economic and security issues. Past experience shows us that people's collective behaviour returns to normal not long after a threat or danger is removed. The tourism industry is resilient and operators are creative in responding to the needs and concerns of guests.[14] With this reasoning, one could argue that tourism in cities will recover fairly quickly once the pandemic is over. On the other hand, the extent in time and space of the pandemic is unprecedented. It has impacted all countries in Europe and virtually all corners of the world. The crisis started in March 2020 and now a year later, at the time of writing, it is unclear when it will really be over. It has caused severe harm to the economies of many countries. These circumstances suggest that the change in people's behaviour will be long lasting and therefore the tourism industry will need many years to return to the pre-pandemic tourist numbers.

As and when concerns about the pandemic recede, another global issue will take centre stage – the global warming that is being caused by the excessively high levels of carbon emissions. There is increasing evidence that global warming is causing extreme weather in different parts of the worlds and this is imposing significant social and financial costs to many cities, especially to those that are most economically vulnerable. The science has long insisted that the global action of countries is needed to reduce carbon emissions. One hundred and ninety countries, including EU member states, signed the 2015 Paris Agreement. This sets out a global framework to avoid dangerous climate change by reducing carbon emissions and limiting the increase in average global temperatures to well below 2°C. In addition, the EU has established a procedure by which member states are required to prepare and implement National Energy and Climate Plans.[15]

The debate on climate change is also relevant to tourism and to post-pandemic tourism recovery. Some academics and environmentalists argue that the pandemic crisis offers the international community an unexpected opportunity considering that human activities need to change if nations are to avoid the worst effects of human-induced climate change. Rather than trying to return to business as usual, the COVID-19 pandemic challenges countries and cities to think about the type of consumption that underpins the unsustainable ways of the travel industry. Even if commercial aviation accounts for 'only' about 2.4 per cent of all emissions from fossil-fuel use, air travel often features in debates on carbon emission reductions.[16] This links into another debate that began in recent years with many concerned people arguing for degrowth strategies especially in destinations suffering from 'overtourism'. Degrowth would address the need to curb what until recently appeared to be the unchecked and unsustainable growth of international air travel.[17]

This debate is relevant at different levels of society starting from personal choices made by individuals and families. People may opt for holidays that use land-based travel rather than air flights. Alternatively they may opt for one long holiday in their favourite Mediterranean resort, rather than go for two shorter ones. During the pandemic, some countries, such as Slovenia, offered staycation vouchers to their own nationals to encourage them to take their holidays within their own country.[18] Even if costly this is an option that countries may adopt after the effects of the pandemic recede.

The Future of Tourism Coalition is an initiative of six international NGOs that have come together to take a stand and appeal for change. They argue that the pandemic crisis has presented an opportunity to the world to choose a more sustainable future for tourism. They set out thirteen guiding principles and urge worldwide commitment to them. One of the guiding principles is to 'mitigate climate impacts'. The Coalition makes several recommendations relating to climate change and tourism,[19] including 'mitigating climate impacts is a matter of balance. Lessening frivolous flying, investing in a balanced carbon offset program, making sure that when you do take a flight, you make it count – e.g. taking fewer, longer holidays'.

The climate change debate also requires cities, countries and the international community to make difficult choices. Some scientists and academics argue that there should be restrictions on aviation to slow down the ever-increasing number of flights. Post-pandemic, pressures will intensify on national governments and the international community to take measures to limit the carbon emissions from air travel. Some countries and business sectors will resist such a move. Economic activity in many cities relies on tourism and air travel, particularly cities and resorts in South Europe and along the northern shore of the Mediterranean. The argument is that there cannot be environmental sustainability if there is no economic sustainability and there cannot be economic sustainability unless there are good air travel connections with the rest of Europe. The argument can be taken further to include Mediterranean cities and resorts along the North African coasts. Where the political climate permits, it is important for North African countries to develop their tourism industry as a means of reinvigorating their economies. This requires air travel. Healthier economies will enhance the lives of residents of North African cities and also stem the flow of migrants across the Mediterranean.

*Chapter 13*

# Conclusion

Cities compete with one another nationally and internationally. To remain competitive, cities adopt strategies that make use of their own innate resources – their histories, spaces, creative energy and talents. They constantly seek to redefine their economic roles as old functions are lost and new functions are sought to take their place. Cities compete to create employment for their residents and they do so by generating inward investment and attracting tourists, to fill their hotels, conference centres, stadia and museums.

Cities are in constant change due to a range of social and economic forces. They seek to guide these changes in a manner that will enhance their competiveness. Tourism is one of many social and economic forces that shape the urban environment. This comes about mostly from projects, both public and private, that are developed to meet demand of tourists as well as locals. Public authorities invest in the public spaces and the infrastructure of the city to upgrade the city environment to make it a pleasant place to live and visit. Such investments also enable the city to cope with actual and potential increases in visitor numbers. Growth in tourism motivates the private sector to invest in the facilities which tourists use such as hotels, catering outlets, visitor attractions and so on.

Same as for any other human activity, most tourism and leisure activities take place in buildings and urban spaces, thus making architecture and the built environment an integral part of such activities. Cities are important to tourism and tourism is important to cities.[1] The physicality of the city is expressed in its architecture. In recent decades, the role of architecture has taken a turn. There are many examples of contemporary architecture across Europe where architecture does not simply provide the 'container' within which tourism activity takes place but the architecture itself becomes the object of touristic interest.

Tourists visit cities for many reasons. In broad terms, travel can be categorised into business and leisure. More specifically, reasons for business travel could be to attend a conference, business meeting or major exhibition. The motivation for leisure travel could be sightseeing, shopping, visiting friends and relatives or simply to spend time relaxing away from mundane daily chores. People who visit a city for business will spend at least some time engaging in leisure activities, enjoying the sights and the experiences that the city has to offer.

More and more people are engaging in leisure and cultural activities in their own city. They behave like tourists in their own city. This comes about because of changing lifestyles and greater affluence. Normal city life sees residents going to concerts or sports events, visiting museums, walking along promenades or socialising at the nearest bar or restaurant. In some parts of the city, the boundaries between tourists and city residents are increasingly blurred, with the tourism activity becoming indistinguishable from the normal daily city life. This is more so in larger cities that easily absorb large numbers of tourists.

The study of tourism is the study of people who are away from their normal environment. It is also the study of the facilities and services that tourists make use of and the impacts that come about from the tourism activity. It involves the motivations and experiences of the tourists, the expectations and adjustments made by the residents of the destination and the roles of those agencies that intercede between them.[2] The study of urban tourism is about tourism in the context of cities. The discussion of urban tourism in this book focuses on two particular aspects. The first is the various facilities that the tourist makes use of at a city destination while the second refers to the nature of the tourist experience of the cities.

A tourist at a city destination makes use of a range of resources. The overall experience of the destination is the outcome of a combination of attractions, transport, accommodation, entertainment and other services, with the various resources and services being supplied by different entities. The fragmented nature of supply is a challenge for a destination management to coordinate across the various sub-sectors of the tourism industry.

City authorities have a particular important role in tourism product development within their overall economic and social objectives. They invest in public spaces and are responsible for their appropriate management. Moreover, city authorities have the responsibility to create the right financial and social conditions for the private sector to invest and develop commercial facilities and services; for tourists as well as locals. There are many companies and small businesses operating in tourism, hospitality and leisure in a city. These are the backbone of a city's tourism product because they produce the services that tourists consume at the destination. The importance of small businesses to tourism cannot be understated. On the other hand, much of the quality of the visitor experience stems from non-commercial activity, such as enjoying the climate and scenery, mixing socially with local inhabitants and strolling in public spaces.[3]

In the years prior to 2020 many countries across Europe saw a remarkable increase in tourism numbers. The tourism industry greatly benefitted from technological advances that led to cheaper airfares and easier means for people to plan and book their own travel. The rapid growth trends came to an abrupt halt with the COVID pandemic. The impact on the tourism and leisure industries has been devastating with tourism activity in 2020 being reduced to a fifth or less to what it was before. The pandemic has affected all sections of society, in their health and their finances, in every country and over an extended period of time. During the pandemic, national governments had to handle a situation made exceedingly complex because of the overlapping of health, humanitarian and economic issues. The uncertainties made the situation even more complex and difficult. It was not possible to forecast the spread of the virus and the efficacy or otherwise of the lockdowns and the various restrictive measures. Nor was it possible to forecast the extent to which the vaccine would eradicate the virus, more so because various virus mutations were developing. On tourism, looking forward to 2022 and beyond, the two big questions are: Will tourism post-pandemic be any different to what it was prior to 2020? When will tourism numbers recover to levels similar to those before the pandemic? Some reflections on these questions and on the pandemic impact on tourism are provided in chapter 12.

Tourism in cities is characterised by a variety of activities, experiences and facilities that are offered to visitors. Walking is an activity to which the tourist will dedicate significant time during a leisure visit to a city. For many tourists, much of that walking time is spent exploring the historic core of the city and hence the importance of historic areas for tourism. Apart from having a cluster of the city's more important attractions, the historic area is often an attraction in its own right.

Visitor attractions have a crucial role in the development and success of tourism destinations. There are several tourism facilities and resources that must be present for a city to be effective as a tourism destination, but it is visitor attractions that play the most important role.

Some cities seek to use iconic architecture to enhance their attractiveness in the context of increased competition between cities. Also, in this context, many cities are resorting to events as a strategy to reposition and differentiate themselves from other cities. Major events are accompanied by major investment in event venues and in urban infrastructure. The city often serves as a backdrop to the urban festival as activities and performances are staged in the city's urban spaces.

# Appendices

# Appendix A
## *World Heritage Site Criteria for Cultural Properties*

World Heritage Sites have to satisfy at least one of ten criteria to demonstrate that they have outstanding universal value to humanity that must be protected for future generations.

There are *six* criteria that are relevant to cultural properties. These are as follows:

i) to represent a masterpiece of human creative genius
ii) to exhibit an important interchange of human values, over a span of time or within a cultural area of the world, on developments in architecture or technology, monumental arts, town-planning or landscape design
iii) to bear a unique or at least exceptional testimony to a cultural tradition or to a civilization which is living or which has disappeared
iv) to be an outstanding example of a type of building, architectural or technological ensemble or landscape which illustrates (a) significant stage(s) in human history
v) to be an outstanding example of a traditional human settlement, land-use or sea-use which is representative of a culture (or cultures), or human interaction with the environment especially when it has become vulnerable under the impact of irreversible change
vi) to be directly or tangibly associated with events or living traditions, with ideas or with beliefs, with artistic and literary works of outstanding universal significance. (The Committee considers that this criterion should preferably be used in conjunction with other criteria).

# Appendix B

## Historic City Areas Inscribed as World Heritage Sites

This appendix includes urban areas that have been inscribed as World Heritage Sites (WHS). In most cases, these are the urban cores of towns or cities with a history of hundreds of years. In some cases, an entire town/mall city is inscribed. Note that WHS that are an individual building or small group of buildings are *not* included in this appendix.

The descriptive text is sourced verbatim from the United Nations Educational, Scientific, and Cultural Organization's (UNESCO) official website (http://whc.unesco.org/en/list/). For more information on any of the inscribed urban areas, refer to UNESCO's website. Each listing includes the city and country, the official name of the property, the year of inscription, the area of the property in hectares, and criteria for WHS inscription.

| Amsterdam, Netherlands | Seventeenth-Century Canal Ring Area of Amsterdam inside the Singelgracht | 2010 | 198 ha | Criteria (i) (ii) (iv) |
|---|---|---|---|---|

The historic urban ensemble of the canal district of Amsterdam was a project for a new 'port city' built at the end of the sixteenth and beginning of the seventeenth centuries. It comprises a network of canals to the west and south of the historic old town and the medieval port that encircled the old town and was accompanied by the repositioning inland of the city's fortified boundaries, the Singelgracht. This was a long-term programme that involved extending the city by draining the swampland, using a system of canals in concentric arcs and filling in the intermediate spaces. These spaces allowed the development of a homogeneous urban ensemble, including gabled houses and numerous monuments. This urban extension was the largest and most homogeneous of its time. It was a model of large-scale town planning, and served as a reference throughout the world until the nineteenth century.

## Appendix B

**Arles, France** — *Arles, Roman and Romanesque Monuments* — 1981 — 65 ha — Criteria (ii) (iv)

Arles is a good example of the adaptation of an ancient city to medieval European civilisation. It has some impressive Roman monuments, of which the earliest – the arena, the Roman theatre and the cryptoporticus (subterranean galleries) – date back to the first century B.C. During the fourth century, Arles experienced a second Golden Age, as attested by the baths of Constantine and the necropolis of Alyscamps. In the eleventh and twelfth centuries, Arles once again became one of the most attractive cities in the Mediterranean. Within the city walls, Saint-Trophime, with its cloister, is one of Provence's major Romanesque monuments.

**Avila, Spain** — *Old Town of Ávila with Its Extra-Muros Churches* — 1985 — 36 ha — Criteria (iii) (iv)

Founded in the eleventh century to protect the Spanish territories from the Moors, this 'City of Saints and Stones', the birthplace of St. Teresa and the burial place of the Grand Inquisitor Torquemada, has kept its medieval austerity. This purity of form can still be seen in the Gothic cathedral and the fortifications which, with their eighty-two semicircular towers and nine gates, are the most complete in Spain.

**Bamberg, Germany** — *Town of Bamberg* — 1993 — 142 ha — Criteria (ii) (iv)

From the tenth century onwards, this town became an important link with the Slav peoples, especially those of Poland and Pomerania. During its period of greatest prosperity, from the twelfth century onwards, the architecture of Bamberg strongly influenced northern Germany and Hungary. In the late eighteenth century, it was the centre of the Enlightenment in southern Germany, with eminent philosophers and writers such as Hegel and Hoffmann living there.

**Banská Štiavnica, Slovakia** — *Historic Town of Banská Štiavnica and the Technical Monuments in Its Vicinity* — 1993 — 21 ha — Criteria (iv) (v)

Over the centuries, the town of Banská Štiavnica was visited by many outstanding engineers and scientists who contributed to its fame. The old medieval mining centre grew into a town with Renaissance palaces, sixteenth-century churches, elegant squares and castles. The urban centre blends into the surrounding landscape, which contains vital relics of the mining and metallurgical activities of the past.

**Bardejov, Slovakia** — *Bardejov Town Conservation Reserve* — 2000 — 24 ha — Criteria (iii) (iv)

Bardejov is a small but exceptionally complete and well-preserved example of a fortified medieval town, which typifies the urbanisation in this region. Among other remarkable features, it also contains a small Jewish quarter around a fine eighteenth-century synagogue.

| Berat and Gjirokastra, Albania | *Historic Centres of Berat and Gjirokastra* | 2008 | 59 ha | Criteria (iii) (iv) |

Berat and Gjirokastra are inscribed as rare examples of an architectural character typical of the Ottoman period. Located in central Albania, Berat bears witness to the coexistence of various religious and cultural communities down the centuries. It features a castle, locally known as the Kala, most of which was built in the thirteenth century, although its origins date back to the fourth century B.C. The citadel area numbers many Byzantine churches, mainly from the thirteenth century, as well as several mosques built under the Ottoman era which began in 1417. Gjirokastra, in the Drinos river valley in southern Albania, features a series of outstanding two-story houses which were developed in the seventeenth century. The town also retains a bazaar, an eighteenth-century mosque and two churches of the same period.

| Berne, Switzerland | *Old City of Berne* | 1983 | 85 ha | Criteria (iii) |

Founded in the twelfth century on a hill site surrounded by the Aare River, Berne developed over the centuries in line with an exceptionally coherent planning concept. The buildings in the Old City, dating from a variety of periods, include fifteenth-century arcades and sixteenth-century fountains. Most of the medieval town was restored in the eighteenth century, but it has retained its original character.

| Bordeaux, France | *Bordeaux, Port of the Moon* | 2007 | 1,731 ha | Criteria (ii) (iv) |

The port city of Bordeaux in southwest France is inscribed as an inhabited historic city, an outstanding urban and architectural ensemble, created in the age of the Enlightenment, whose values continued up to the first half of the twentieth century, with more protected buildings than any other French city except Paris.

| Brugges, Belgium | *Historic Centre of Brugge* | 2000 | 410 ha | Criteria (ii) (iv) (vi) |

Brugge is an outstanding example of a medieval historic settlement, which has maintained its historic fabric as this has evolved over the centuries, and where original Gothic constructions form part of the town's identity. As one of the commercial and cultural capitals of Europe, Brugge developed cultural links to different parts of the world. It is closely associated with the school of Flemish Primitive painting.

| Budapest, Hungary | *Budapest, including the Banks of the Danube, the Buda Castle Quarter and Andrássy Avenue* | 1987 | 474 ha | Criteria (ii) (iv) |

This site has the remains of monuments such as the Roman city of Aquincum and the Gothic castle of Buda, which have had a considerable influence on the architecture of various periods. It is one of the world's outstanding urban landscapes and illustrates the great periods in the history of the Hungarian capital.

## Appendix B

| Český Krumlov, Czech Republic | *Historic Centre of Český Krumlov* | 1992 | 52 ha | Criteria (iv) |

Situated on the banks of the Vltava river, the town was built around a thirteenth-century castle with Gothic, Renaissance and Baroque elements. It is an outstanding example of a small central European medieval town whose architectural heritage has remained intact thanks to its peaceful evolution over more than five centuries.

| Christiansfeld, Denmark | *Christiansfeld, a Moravian Church Settlement* | 2015 | 21 ha | Criteria (iii) (iv) |

Founded in 1773 in South Jutland, the site is an example of a planned settlement of the Moravian Church, a Lutheran free congregation centred in Herrnhut, Saxony. The town was planned to represent the Protestant urban ideal, constructed around a central Church square. The architecture is homogenous and unadorned, with one and two-storey buildings in yellow brick with red tile roofs. The democratic organisation of the Moravian Church, with its pioneering egalitarian philosophy, is expressed in its humanistic town planning. The buildings are still in use and many are still owned by the local Moravian Church community.

| Cordoba, Spain | *Historic Centre of Cordoba* | 1984 | n/a | Criteria (i) (ii) (iii) (iv) |

Cordoba's period of greatest glory began in the eighth century after the Moorish conquest, when some 300 mosques and innumerable palaces and public buildings were built to rival the splendours of Constantinople, Damascus and Baghdad. In the thirteenth century, under Ferdinand III, the Saint, Cordoba's Great Mosque was turned into a cathedral and new defensive structures, particularly the Alcázar de los Reyes Cristianos and the Torre Fortaleza de la Calahorra, were erected.

| Corfu, Greece | *Old Town of Corfu* | 2007 | 70 ha | Criteria (iv) |

The Old Town of Corfu, on the Island of Corfu off the western coasts of Albania and Greece, is located in a strategic position at the entrance of the Adriatic Sea, and has its roots in the eighth century B.C. The three forts of the town, designed by renowned Venetian engineers, were used for four centuries to defend the maritime trading interests of the Republic of Venice against the Ottoman Empire. In the course of time, the forts were repaired and partly rebuilt several times, more recently under British rule in the nineteenth century. The mainly neoclassical housing stock of the Old Town is partly from the Venetian period, partly of later construction, notably the nineteenth century. As a fortified Mediterranean port, Corfu's urban and port ensemble is notable for its high level of integrity and authenticity.

| Dubrovnik, Croatia | *Old City of Dubrovnik* | 1979 | 97 ha | Criteria (i) (iii) (iv) |

The 'Pearl of the Adriatic', situated on the Dalmatian coast, became an important Mediterranean sea power from the thirteenth century onwards. Although severely damaged by an earthquake in 1667, Dubrovnik managed to preserve its beautiful Gothic, Renaissance and Baroque churches, monasteries, palaces and fountains. Damaged again in the 1990s by armed conflict, it is now the focus of a major restoration programme co-ordinated by UNESCO.

| | | | | |
|---|---|---|---|---|
| **Edinburgh, Scotland** | *Old and New Towns of Edinburgh* | 1995 | n/a | Criteria (ii) (iv) |

Edinburgh has been the Scottish capital since the fifteenth century. It has two distinct areas: the Old Town, dominated by a medieval fortress, and the neoclassical New Town, whose development from the eighteenth century onwards had a far-reaching influence on European urban planning. The harmonious juxtaposition of these two contrasting historic areas, each with many important buildings, is what gives the city its unique character.

| | | | | |
|---|---|---|---|---|
| **Évora, Portugal** | *Historic Centre of Évora* | 1986 | n/a | Criteria (ii) (iv) |

This museum-city, whose roots go back to Roman times, reached its golden age in the fifteenth century, when it became the residence of the Portuguese kings. Its unique quality stems from the whitewashed houses decorated with azulejos and wrought-iron balconies dating from the sixteenth to the eighteenth centuries. Its monuments had a profound influence on Portuguese architecture in Brazil.

| | | | | |
|---|---|---|---|---|
| **Ferrara, Italy** | *Ferrara, City of the Renaissance, and its Po Delta* | 1995 | 46,712 ha | Criteria (ii) (iii) (iv) (v) (vi) |

Ferrara, which grew up around a ford over the River Po, became an intellectual and artistic centre that attracted the greatest minds of the Italian Renaissance in the fifteenth and sixteenth centuries. Here, Piero della Francesca, Jacopo Bellini and Andrea Mantegna decorated the palaces of the House of Este. The humanist concept of the 'ideal city' came to life here in the neighbourhoods built from 1492 onwards by Biagio Rossetti according to the new principles of perspective. The completion of this project marked the birth of modern town planning and influenced its subsequent development.

| | | | | |
|---|---|---|---|---|
| **Florence, Italy** | *Historic Centre of Florence* | 1982 | 505 ha | Criteria (i) (ii) (iii) (iv) (vi) |

Built on the site of an Etruscan settlement, Florence, the symbol of the Renaissance, rose to economic and cultural pre-eminence under the Medici in the fifteenth and sixteenth centuries. Its 600 years of extraordinary artistic activity can be seen above all in the thirteenth-century cathedral (Santa Maria del Fiore), the Church of Santa Croce, the Uffizi and the Pitti Palace, the work of great masters such as Giotto, Brunelleschi, Botticelli and Michelangelo.

| | | | | |
|---|---|---|---|---|
| **Graz, Austria** | *City of Graz – Historic Centre and Schloss Eggenberg* | 1999 | n/a | Criteria (ii) (iv) |

The City of Graz – Historic Centre and Schloss Eggenberg bear witness to an exemplary model of the living heritage of a central European urban complex influenced by the secular presence of the Habsburgs and the cultural and artistic role played by the main aristocratic families. They are a harmonious blend of the architectural styles and artistic movements that have succeeded each other from the Middle Ages until the eighteenth century, from the many neighbouring regions of Central and Mediterranean Europe. They embody a diversified and highly comprehensive ensemble of architectural, decorative and landscape examples of these interchanges of influence.

**Greenwich,**   *Maritime Greenwich*   1997   109 ha   Criteria (i) (ii)
**UK**                                                                (iv) (vi)

The ensemble of buildings at Greenwich, an outlying district of London, and the park in which they are set, symbolise English artistic and scientific endeavour in the seventeenth and eighteenth centuries. The Queen's House (by Inigo Jones) was the first Palladian building in England, while the complex that was until recently the Royal Naval College was designed by Christopher Wren. The park, laid out on the basis of an original design by André Le Nôtre, contains the Old Royal Observatory, the work of Wren and the scientist Robert Hooke.

**Hamburg,**   *Speicherstadt and Kontorhaus*   2015   26 ha   Criteria (iv)
**Germany**    *Districts with Chilehaus*

Speicherstadt and the adjacent Kontorhaus districts are two densely built urban areas in the centre of the port city of Hamburg. Speicherstadt, originally developed on a group of narrow islands in the Elbe River between 1885 and 1927, was partly rebuilt from 1949 to 1967. It is one of the largest coherent historic ensembles of port warehouses in the world (300,000 m$^2$). It includes fifteen very large warehouse blocks as well as six ancillary buildings and a connecting network of short canals. Adjacent to the modernist Chilehaus office building, the Kontorhaus district is an area of over five hectares featuring six very large office complexes built from the 1920s to the 1940s to house port-related businesses. The complex exemplifies the effects of the rapid growth in international trade in the late nineteenth and early twentieth centuries.

**Karlskrona,**   *Naval Port of Karlskrona*   1998   320 ha   Criteria (ii)
**Sweden**                                                            (iv)

Karlskrona is an outstanding example of a late-seventeenth-century European planned naval city. The original plan and many of the buildings have survived intact, along with installations that illustrate its subsequent development up to the present day.

**Kraków,**   *Historic Centre of Kraków*   1978   150 ha   Criteria (iv)
**Poland**

The Historic Centre of Kraków, the former capital of Poland, is situated at the foot of the Royal Wawel Castle. The thirteenth-century merchants' town has Europe's largest market square and numerous historical houses, palaces and churches with their magnificent interiors. Further evidence of the town's fascinating history is provided by the remnants of the fourteenth-century fortifications and the medieval site of Kazimierz with its ancient synagogues in the southern part of town, Jagellonian University and the Gothic cathedral where the kings of Poland were buried.

**Kutná Hora,**   *Kutná Hora: Historical Town*   1995   62 ha   Criteria (ii)
**Czech Rep.**   *Centre with the Church of St*                          (iv)
                 *Barbara and the Cathedral of*
                 *Our Lady at Sedlec*

Kutná Hora developed as a result of the exploitation of the silver mines. In the fourteenth century, it became a royal city endowed with monuments that symbolised its prosperity. The Church of St Barbara, a jewel of the late Gothic period, and the Cathedral of Our Lady at Sedlec, which was restored in line with the Baroque taste of the early eighteenth century, were to influence the architecture of central Europe. These masterpieces today form part of a well-preserved medieval urban fabric with some particularly fine private dwellings.

| Le Havre, France | *Le Havre, the City Rebuilt by Auguste Perret* | 2005 | 133 ha | Criteria (ii) (iv) |

The city of Le Havre, on the English Channel in Normandy, was severely bombed during the Second World War. The destroyed area was rebuilt according to the plan of a team headed by Auguste Perret, from 1945 to 1964. The site forms the administrative, commercial and cultural centre of Le Havre. Le Havre is exceptional among many reconstructed cities for its unity and integrity. It combines a reflection of the earlier pattern of the town and its extant historic structures with the new ideas of town planning and construction technology. It is an outstanding post-war example of urban planning and architecture based on the unity of methodology and the use of prefabrication, the systematic utilisation of a modular grid and the innovative exploitation of the potential of concrete.

| Liverpool, UK | *Liverpool – Maritime Mercantile City* | 2004 | 136 ha | Criteria (ii) (iii) (iv) |

Six areas in the historic centre and docklands of the maritime mercantile City of Liverpool bear witness to the development of one of the world's major trading centres in the eighteenth and nineteenth centuries. Liverpool was a pioneer in the development of modern dock technology, transport systems and port management. The listed sites feature a great number of significant commercial, civic and public buildings, including St. George's Plateau.

| Lubeck, Germany | *Hanseatic City of Lübeck* | 1987 | 81 ha | Criteria (iv) |

Lübeck – the former capital and Queen City of the Hanseatic League – was founded in the twelfth century and prospered until the sixteenth century as the major trading centre for Northern Europe. It has remained a centre for maritime commerce to this day, particularly with the Nordic countries. Despite the damage it suffered during World War II, the basic structure of the old city, consisting mainly of fifteenth- and sixteenth-century patrician residences, public monuments (the famous Holstentor brick gate), churches and salt storehouses, remains unaltered.

| Luxembourg | *City of Luxembourg: Its Old Quarters and Fortifications* | 1994 | 30 ha | Criteria (iv) |

Because of its strategic position, Luxembourg was, from the sixteenth century until 1867, when its walls were dismantled, one of Europe's greatest fortified sites. It was repeatedly reinforced as it passed from one great European power to another: the Holy Roman Emperors, the House of Burgundy, the Habsburgs, the French and Spanish kings and finally the Prussians. Until their partial demolition, the fortifications were a fine example of military architecture spanning several centuries.

| Lyon, France | *Historic Site of Lyon* | 1998 | 427 ha | Criteria (ii) (iv) |

The long history of Lyon, which was founded by the Romans in the first century B.C. as the capital of the Three Gauls and has continued to play a major role in Europe's political, cultural and economic development ever since, is vividly illustrated by its urban fabric and the many fine historic buildings from all periods.

## Appendix B

| | | | | |
|---|---|---|---|---|
| Naples, Italy | *Historic Centre of Naples* | 1995 | 1,021 ha | Criteria (ii) (iv) |

From the Neapolis founded by Greek settlers in 470 B.C. to the city of today, Naples has retained the imprint of the successive cultures that emerged in Europe and the Mediterranean basin. This makes it a unique site, with a wealth of outstanding monuments such as the Church of Santa Chiara and the Castel Nuovo.

| | | | | |
|---|---|---|---|---|
| Oporto, Portugal | *Historic Centre of Oporto, Luiz I Bridge and Monastery of Serra do Pilar* | 1996 | | Criteria (iv) |

The city of Oporto, built along the hillsides overlooking the mouth of the Douro river, is an outstanding urban landscape with a 2,000-year history. Its continuous growth, linked to the sea (the Romans gave it the name Portus, or port), can be seen in the many and varied monuments, from the cathedral with its Romanesque choir, to the neoclassical Stock Exchange and the typically Portuguese Manueline-style Church of Santa Clara.

| | | | | |
|---|---|---|---|---|
| Paris, France | *Paris, Banks of the Seine* | 1991 | 365 ha | Criteria (i) (ii) (iv) |

From the Louvre to the Eiffel Tower, from the Place de la Concorde to the Grand and Petit Palais, the evolution of Paris and its history can be seen from the River Seine. The Cathedral of Notre-Dame and the Sainte Chapelle are architectural masterpieces while Haussmann's wide squares and boulevards influenced late nineteenth- and twentieth-century town planning the world over.

| | | | | |
|---|---|---|---|---|
| Prague, Czech Republic | *Historic Centre of Prague* | 1992 | 1,106 ha | Criteria (ii) (iv) (vi) |

Built between the eleventh and eighteenth centuries, the Old Town, the Lesser Town and the New Town speak of the great architectural and cultural influence enjoyed by this city since the Middle Ages. The many magnificent monuments, such as Hradcany Castle, St. Vitus Cathedral, Charles Bridge and numerous churches and palaces, built mostly in the fourteenth century under the Holy Roman Emperor, Charles IV.

| | | | | |
|---|---|---|---|---|
| Provins, France | *Provins, Town of Medieval Fairs* | 2001 | 108 ha | Criteria (ii) (iv) |

The fortified medieval town of Provins is situated in the former territory of the powerful Counts of Champagne. It bears witness to early developments in the organisation of international trading fairs and the wool industry. The urban structure of Provins, which was built specifically to host the fairs and related activities, has been well preserved.

| | | | | |
|---|---|---|---|---|
| Quedlinberg, Germany | *Collegiate Church, Castle and Old Town of Quedlinburg* | 1994 | 270 ha | Criteria (iv) |

Quedlinburg, in the Land of Sachsen-Anhalt, was a capital of the East Franconian German Empire at the time of the Saxonian-Ottonian ruling dynasty. It has been a prosperous trading town since the Middle Ages. The number and high quality of the timber-framed buildings make Quedlinburg an exceptional example of a medieval European town. The Collegiate Church of St. Servatius is one of the masterpieces of Romanesque architecture.

| Rauma, Finland | *Old Rauma* | 1991 | 29 ha | Criteria (iv) (v) |

Situated on the Gulf of Botnia, Rauma is one of the oldest harbours in Finland. Built around a Franciscan monastery, where the mid-fifteenth-century Holy Cross Church still stands, it is an outstanding example of an old Nordic city constructed in wood. Although ravaged by fire in the late seventeenth century, it has preserved its ancient vernacular architectural heritage.

| Regensburg, Germany | *Old Town of Regensburg with Stadtamhof* | 2006 | 183 ha | Criteria (ii) (iii) (iv) |

Located on the Danube River in Bavaria, this medieval town contains many buildings of exceptional quality that testify to its history as a trading centre and to its influence on the region from the ninth century. A notable number of historic structures span some two millennia and include ancient Roman, Romanesque and Gothic buildings. Regensburg's eleventh- to thirteenth-century architecture – including the market, city hall and cathedral – still defines the character of the town marked by tall buildings, dark and narrow lanes, and strong fortifications. The buildings include medieval patrician houses and towers, a large number of churches and monastic ensembles as well as the twelfth-century Old Bridge. The town is also remarkable for the vestiges testifying to its rich history as one of the centres of the Holy Roman Empire that turned to Protestantism.

| Rhodes, Greece | *Medieval City of Rhodes* | 1980 | 66 ha | Criteria (ii) (iv) (v) |

The Order of St John of Jerusalem occupied Rhodes from 1309 to 1523 and set about transforming the city into a stronghold. It subsequently came under Turkish and Italian rule. With the Palace of the Grand Masters, the Great Hospital and the Street of the Knights, the Upper Town is one of the most beautiful urban ensembles of the Gothic period. In the Lower Town, Gothic architecture coexists with mosques, public baths and other buildings dating from the Ottoman period.

| Riga, Latvia | *Historic Centre of Riga* | 1997 | 438 ha | Criteria (i) (ii) |

Riga was a major centre of the Hanseatic League, deriving its prosperity in the thirteenth to fifteenth centuries from the trade with Central and Eastern Europe. The urban fabric of its medieval centre reflects this prosperity, though most of the earliest buildings were destroyed by fire or war. Riga became an important economic centre in the nineteenth century, when the suburbs surrounding the medieval town were laid out, first with imposing wooden buildings in neoclassical style and then in Jugendstil. It is generally recognised that Riga has the finest collection of art nouveau buildings in Europe.

| Rome, Italy | *Historic Centre of Rome, the Properties of the Holy See in That City Enjoying Extraterritorial Rights and San Paolo Fuori le Mura* | 1980 | | Criteria (i) (ii) (iii) (iv) (vi) |

Founded, according to legend, by Romulus and Remus in 753 B.C., Rome was first the centre of the Roman Republic, then of the Roman Empire, and it became the capital of the Christian world in the fourth century. The World Heritage site, extended in 1990 to the walls of Urban VIII, includes some of the major monuments of antiquity such as the Forums, the Mausoleum of Augustus, the Mausoleum of Hadrian, the Pantheon, Trajan's Column and the Column of Marcus Aurelius, as well as the religious and public buildings of papal Rome.

| Salamanca, Spain | *Old City of Salamanca* | 1988 | 51 ha | Criteria (i) (ii) (iv) |

This ancient university town north-west of Madrid was first conquered by the Carthaginians in the third century B.C. It then became a Roman settlement before being ruled by the Moors until the eleventh century. The university, one of the oldest in Europe, reached its high point during Salamanca's golden age. The city's historic centre has important Romanesque, Gothic, Moorish, Renaissance and Baroque monuments. The Plaza Mayor, with its galleries and arcades, is particularly impressive.

| Salzburg, Austria | *Historic Centre of the City of Salzburg* | 1996 | 236 ha | Criteria (ii) (iv) (vi) |

Salzburg has managed to preserve an extraordinarily rich urban fabric, developed over the period from the Middle Ages to the nineteenth century when it was a city-state ruled by a prince-archbishop. Its flamboyant Gothic art attracted many craftsmen and artists before the city became even better known through the work of the Italian architects Vincenzo Scamozzi and Santini Solari, to whom the centre of Salzburg owes much of its Baroque appearance. This meeting-point of Northern and Southern Europe perhaps sparked the genius of Salzburg's most famous son, Wolfgang Amadeus Mozart, whose name has been associated with the city ever since.

| San Gimignano, Italy | *Historic Centre of San Gimignano* | 1990 | 14 ha | Criteria (i) (iii) (iv) |

'San Gimignano delle belle Torri' is in Tuscany, 56 km south of Florence. It served as an important relay point for pilgrims travelling to or from Rome on the Via Francigena. The patrician families who controlled the town built around seventy-two tower-houses (some as high as 50 m) as symbols of their wealth and power. Although only fourteen have survived, San Gimignano has retained its feudal atmosphere and appearance. The town also has several masterpieces of fourteenth- and fifteenth-century Italian art.

## Historic City Areas

**San Marino**     ***San Marino Historic Centre and***    2008    55 ha    Criteria (iii)
***Mount Titano***

San Marino Historic Centre and Mount Titano covers 55 ha, including Mount Titano and the historic centre of the city which dates back to the foundation of the republic as a city-state in the thirteenth century. San Marino is inscribed as a testimony to the continuity of a free republic since the Middle Ages. The inscribed city centre includes fortification towers, walls, gates and bastions, as well as a neo-classical basilica of the nineteenth century, fourteenth- and sixteenth-century convents, and the Palazzo Publico of the nineteenth century, as well as the eighteenth-century Titano Theatre. The property represents an historical centre still inhabited and preserving all its institutional functions.

**Santiago de**     ***Santiago de Compostela (Old***    1985    108 ha    Criteria (i) (ii)
**Compostela,**     ***Town)***     (vi)
**Spain**

This famous pilgrimage site in north-west Spain became a symbol in the Spanish Christians' struggle against Islam. Destroyed by the Muslims at the end of the tenth century, it was completely rebuilt in the following century. With its Romanesque, Gothic and Baroque buildings, the Old Town of Santiago is one of the world's most beautiful urban areas. The oldest monuments are grouped around the tomb of St James and the cathedral, which contains the remarkable Pórtico de la Gloria.

**Segovia, Spain**     ***Old Town of Segovia and Its***    1985    134 ha    Criteria (i)
***Aqueduct***     (iii) (iv)

The Roman aqueduct of Segovia, probably built c. A.D. 50, is remarkably well preserved. This impressive construction, with its two tiers of arches, forms part of the setting of the magnificent historic city of Segovia. Other important monuments include the Alcázar, begun around the eleventh century, and the sixteenth-century Gothic cathedral.

**Siena, Italy**     ***Historic Centre of Siena***    1995    170 ha    Criteria (i) (ii)
(iv)

Siena is the embodiment of a medieval city. Its inhabitants pursued their rivalry with Florence right into the area of urban planning. Throughout the centuries, they preserved their city's Gothic appearance, acquired between the twelfth and fifteenth centuries. During this period, the work of Duccio, the Lorenzetti brothers and Simone Martini was to influence the course of Italian and, more broadly, European art. The whole city of Siena, built around the Piazza del Campo, was devised as a work of art that blends into the surrounding landscape.

**Stralsund and**     ***Historic Centres of Stralsund***    2002    168 ha    Criteria (ii)
**Wismar,**     ***and Wismar***     (iv)
**Germany**

The medieval towns of Wismar and Stralsund were major trading centres of the Hanseatic League in the fourteenth and fifteenth centuries. In the seventeenth and eighteenth centuries, they became Swedish administrative and defensive centres for the German territories. They contributed to the development of the characteristic building types and techniques of Brick Gothic in the Baltic region, as exemplified in several important brick cathedrals, the Town Hall of Stralsund, and the series of houses for residential, commercial and crafts use, representing its evolution over several centuries.

**Strasbourg, France**   *Strasbourg, Grande-Île and Neustadt*   1988   183 ha   Criteria (ii) (iv)

The initial property, inscribed in 1988 on the World Heritage List, was formed by the Grande-Île, the historic centre of Strasbourg, structured around the cathedral. The extension concerns the Neustadt, new town, designed and built under the German administration (1871–1918). The Neustadt draws the inspiration for its urban layout partially from the Haussmannian model, while adopting an architectural idiom of Germanic inspiration. This dual influence has enabled the creation of an urban space that is specific to Strasbourg, where the perspectives created around the cathedral open to a unified landscape around the rivers and canals.

**Tallinn, Estonia**   *Historic Centre (Old Town) of Tallinn*   1997   113 ha   Criteria (ii) (iv)

The origins of Tallinn date back to the thirteenth century, when a castle was built there by the crusading knights of the Teutonic Order. It developed as a major centre of the Hanseatic League, and its wealth is demonstrated by the opulence of the public buildings (the churches in particular) and the domestic architecture of the merchants' houses, which have survived to a remarkable degree despite the ravages of fire and war in the intervening centuries.

**Telč, Czech Republic**   *Historic Centre of Telč*   1992   36 ha   Criteria (i) (iv)

The houses in Telc, which stands on a hilltop, were originally built of wood. After a fire in the late fourteenth century, the town was rebuilt in stone, surrounded by walls and further strengthened by a network of artificial ponds. The town's Gothic castle was reconstructed in High Gothic style in the late fifteenth century.

**Toledo, Spain**   *Historic City of Toledo*   1986   260 ha   Criteria (i) (ii) (iii) (iv)

Successively a Roman municipium, the capital of the Visigothic Kingdom, a fortress of the Emirate of Cordoba, an outpost of the Christian kingdoms fighting the Moors and, in the sixteenth century, the temporary seat of supreme power under Charles V, Toledo is the repository of more than 2,000 years of history. Its masterpieces are the product of heterogeneous civilisations in an environment where the existence of three major religions—Judaism, Christianity and Islam—was a major factor.

**Toruń, Poland**   *Medieval Town of Toruń*   1997   48 ha   Criteria (ii) (iv)

Torun owes its origins to the Teutonic Order, which built a castle there in the mid-thirteenth century as a base for the conquest and evangelisation of Prussia. It soon developed a commercial role as part of the Hanseatic League. In the Old and New Town, the many imposing public and private buildings from the fourteenth and fifteenth centuries (among them the house of Copernicus) are striking evidence of Torun's importance.

**Urbino, Italy**   *Historic Centre of Urbino*   1998   29 ha   Criteria (ii) (iv)

The small hill town of Urbino, in the Marche, experienced a great cultural flowering in the fifteenth century, attracting artists and scholars from all over Italy and beyond, and influencing cultural developments elsewhere in Europe. Owing to its economic and cultural stagnation from the sixteenth century onwards, it has preserved its Renaissance appearance to a remarkable extent.

## Historic City Areas

**Valletta, Malta** *City of Valletta*    1980    55 ha    Criteria (i) (vi)

The capital of Malta is inextricably linked to the history of the military and charitable Order of St John of Jerusalem. It was ruled successively by the Phoenicians, Greeks, Carthaginians, Romans, Byzantines, Arabs and the Order of the Knights of St John. Valletta's 320 monuments, all within an area of 55 ha, make it one of the most concentrated historic areas in the world.

**Venice, Italy** *Venice and its Lagoon*    1987    70,176 ha    Criteria (i) (ii) (iii) (iv) (v) (vi)

Founded in the fifth century and spread over 118 small islands, Venice became a major maritime power in the tenth century. The entire city is an extraordinary architectural masterpiece in which even the smallest building contains works by some of the world's greatest artists such as Giorgione, Titian, Tintoretto, Veronese and others.

**Verona, Italy** *City of Verona*    2000    444 ha    Criteria (ii) (iv)

The historic city of Verona was founded in the first century B.C. It flourished under the rule of the Scaliger family in the thirteenth and fourteenth centuries and as part of the Republic of Venice from the fifteenth to eighteenth centuries. Verona has preserved a remarkable number of monuments from antiquity and the medieval and Renaissance periods, and represents an outstanding example of a military stronghold.

**Vicenza, Italy** *City of Vicenza and the Palladian Villas of the Veneto*    1994    334 ha    Criteria (i) (ii)

Founded in the second century B.C. in Northern Italy, Vicenza prospered under Venetian rule from the early fifteenth century to the end of the eighteenth century. The work of Andrea Palladio (1508–1580), based on a detailed study of classical Roman architecture, gives the city its unique appearance. Palladio's urban buildings, as well as his villas, scattered throughout the Veneto region, had a decisive influence on the development of architecture. His work inspired a distinct architectural style known as Palladian, which spread to England and other European countries, and also to North America.

**Vienna, Austria** *Historic Centre of Vienna*    2001    371 ha    Criteria (ii) (iv) (vi)

Vienna developed from early Celtic and Roman settlements into a Medieval and Baroque city, the capital of the Austro-Hungarian Empire. It played an essential role as a leading European music centre, from the great age of Viennese Classicism through the early part of the twentieth century. The historic centre of Vienna is rich in architectural ensembles, including Baroque castles and gardens, as well as the late-nineteenth-century Ringstrasse lined with grand buildings, monuments and parks.

**Vilnius, Lithuania** *Vilnius Historic Centre*    1994    352 ha    Criteria (ii) (iv)

Political centre of the Grand Duchy of Lithuania from the thirteenth to the end of the eighteenth century, Vilnius has had a profound influence on the cultural and architectural development of much of Eastern Europe. Despite invasions and partial destruction, it has preserved an impressive complex of Gothic, Renaissance, Baroque and classical buildings as well as its medieval layout and natural setting.

**Visby, Sweden**  *Hanseatic Town of Visby*  1995  105 ha  Criteria (iv) (v)

A former Viking site on the island of Gotland, Visby was the main centre of the Hanseatic League in the Baltic from the twelfth to the fourteenth century. Its thirteenth-century ramparts and more than 200 warehouses and wealthy merchants' dwellings from the same period make it the best-preserved fortified commercial city in Northern Europe.

**Warsaw, Poland**  *Historic Centre of Warsaw*  1980  26 ha  Criteria (ii) (vi)

During the Warsaw Uprising in August 1944, more than 85 per cent of Warsaw's historic centre was destroyed by Nazi troops. After the war, a five-year reconstruction campaign by its citizens resulted in today's meticulous restoration of the Old Town, with its churches, palaces and marketplace. It is an outstanding example of a near-total reconstruction of a span of history covering the thirteenth to the twentieth centuries.

**Zamość, Poland**  *Old City of Zamość*  1992  75 ha  Criteria (iv)

Zamosc was founded in the sixteenth century by the chancellor Jan Zamoysky on the trade route linking Western and Northern Europe with the Black Sea. Modelled on Italian theories of the 'ideal city' and built by the architect Bernando Morando, a native of Padua, Zamosc is a perfect example of a late-sixteenth-century Renaissance town. It has retained its original layout and fortifications and a large number of buildings that combine Italian and central European architectural traditions.

# Notes

## CHAPTER 1

1. Wall, G., & Mathieson, A. (2005) *Tourism: Change, Impacts and Opportunities*. Harlow: Pearson.
2. Ashworth, G., & Page, S. (2011) Urban tourism research: Recent progress and current paradoxes. *Tourism Management* 32, 1–15.
3. Ashworth & Page, 2011.
4. Smith, A. (2019) Conceptualising the expansion of destination London: Some conclusions. In A. Smith and A. Graham (eds.), *Destination London: The Expansion of the Visitor Economy* (pp. 225–236). London: University of Westminster Press.
5. Edwards, D., Griffin, T., & Hayllar, B. (2008) Urban tourism research, developing an agenda. *Annals of Tourism Research* 35(4), 1032–1052.
6. Ashworth, G., & Tunbridge, J. (2000) *The Tourist-Historic Cities: Retrospect and Prospect of Managing the Heritage City*. London: Routledge. Page 58.
7. Edwards et al., 2008.
8. Ashworth & Tunbridge, 2000.
9. Edwards et al., 2008.
10. Wall & Mathieson, 2005.
11. Wall & Mathieson, 2005.
12. Edwards et al., 2008.
13. Edwards et al., 2008
14. Hall, C. M., & Page, S. J. (2014) *The Geography of Tourism and Recreation. Environment, Place and Space* (fourth edition). London: Routledge.
15. Law, C. (2002) *Urban Tourism: The Visitor Economy and the Growth of Large Cities* (second edition). New York: Continuum.
16. Specht, J. (2014) *Architectural Tourism: Building for Urban Travel Destinations*. Wiesbaden: Springer Gabler.
17. Prideaux, B. (2009) *Resort Destinations: Evolution, Management and Development*. London: Routledge. Page 154.

18. Law, 2002.
19. Prideaux, 2009. Page 140.
20. Law, 2002.
21. Law, 2002.
22. Richter, J. (ed.) (2010) *The Tourist City Berlin: Tourism & Architecture*. Switzerland: Braun Publishing.
23. Kolb, B. (2017) *Tourism Marketing for Cities and Towns: Using Social Media and Branding to Attract Tourists*. London: Routledge.
24. Richter, 2010.
25. Richter, 2010.

# CHAPTER 2

1. Prideaux, B. (2009) *Resort Destinations. Evolution, Management and Development*. London: Routledge.
2. Gordon, I., & Buck, N. (2005) Introduction: Cities in the new conventional wisdom. In N. Buck, I. Gordon, A. Harding, & I. Turok (eds.), *Changing Cities; Rethinking Urban Competiveness, Cohesion and Governance* (pp. 1–21). London: Palgrave Macmillan.
3. Hall, T. (1997) *Planning Europe's Capital Cities: Aspects of Nineteenth Century Urban Development*. London: E&FN Spon.
4. Prideaux, 2009.
5. Hall, P. (1998). *Cities in civilization*. London: Weidenfeld and Nicholson.
6. Shoval, N. (2018) Urban planning and tourism in European cities. *Tourism Geographies* 20(3), 371–376.
7. Couch, C., Sykes, O., & Cocks, M. (2013) The changing context of urban regeneration in north west Europe. In M. E. Leary & J. McCarthy (eds.), *The Routledge Companion to Urban Regeneration* (pp. 33–44). New York: Routledge.
8. Tallon, A. (2013) *Urban Regeneration in the UK* (second edition). London: Routledge.
9. Law, C. (2002) *Urban Tourism: The Visitor Economy and the Growth of Large Cities* (second edition). New York: Continuum.
10. Specht, J. (2014) *Architectural Tourism: Building for Urban Travel Destinations*. Wiesbaden: Springer Gabler.
11. Boisen, M., Terlouw, K., Groote, P., & Couwenberg, O. (2018) Reframing place promotion, place marketing, and place branding: Moving beyond conceptual confusion. *Cities* 80, 4–11.
12. Law, C. M. (2000) Regenerating the city centre through leisure and tourism. *Built Environment* 26, 117–129.
13. Pine, B., & Gilmore, J. (1999) *The Experience Economy*. Boston, MA: Harvard Business School Press.
14. Maitland, R. (2006) How can we manage the tourist-historic city? Tourism strategy in Cambridge, UK, 1978–2003. *Tourism Management* 27(6), 1262–1273.
15. Gordon & Buck, 2005.
16. Tallon, 2013.

17. Colomb, C. (2012) *Staging the New Berlin: Place Marketing and the Politics of Urban Reinvention Post-1989*. London: Routledge.

18. Ponzini, D., Fotev, S., & Mavaacchio, F. (2016), Place making or place faking? The paradoxical effects of transnational circulation of architectural and urban development projects. In A. Russo & G. Richards (eds.), *Reinventing the Local in Tourism: Producing, Consuming and Negotiating Place* (pp. 153–170). Bristol: Channel View Publications.

19. Gordon & Buck, 2005.

20. Law, 2002.

21. Holcomb, B. (1999) Marketing cities for tourism. In D. R. Judd & S. S. Fainstein (eds.), *The Tourist City*. New Haven, CT: Yale University Press.

22. Campelo, A. (2017) The state of the art: From country-of origin to strategies for economic development. In A. Campelo (ed.), *Handbook on Place Branding and Marketing* (pp. 3–21). Elgaronline.

23. Colomb, 2012.

24. Holcomb, 1999.

25. Lorentzen, A. (2009) Cities in the experience economy. *European Planning Studies* 17(6), 829–845.

26. Selby, M. (2004) *Understanding Urban Tourism: Image, Culture and Experience*. London: I. B. Tauris.

27. Tallon, 2013.

28. Colomb, 2012.

29. Colomb, 2012.

30. Boisen et al., 2018.

31. Boisen et al., 2018.

32. Colomb, 2012.

33. Holcomb, 1999.

34. Holcomb, 1999.

35. Kolb, B. (2017) *Tourism Marketing for Cities and Towns: Using Social Media and Branding to Attract Tourists*. London: Routledge.

36. Kolb, 2017.

37. Kolb, 2017.

38. Kolb, 2017.

39. Colomb, 2012.

40. Campelo, 2017.

41. Kolb, 2017.

42. Colomb, 2012.

43. Boisen et al., 2018.

44. Law, 2002.

45. Lorentzen, 2009.

46. Trueman, M., Cook, D., & Cornelius, N. (2008) Creative dimensions for branding and regeneration: Overcoming negative perceptions of a city. *Place Branding and Public Diplomacy* 4(1), 29–44.

47. Campelo, 2017.

48. Spirou, C. (2011) *Urban Tourism and Urban Change: Cities in a Global Economy*. New York: Routledge.

49. Ritchie, J., & Crouch, G. (2003). *The Competitive Destination: A Sustainable Tourism Perspective*. Cambridge, MA: CABI Publishing.
50. Ritchie & Crouch, 2003.
51. Ritchie & Crouch, 2003. Page 23.
52. The relevance of contemporary architecture and iconic buildings is discussed in chapter 9.
53. The relevance of events to cities is discussed in chapter 10.
54. Urban spaces and their relevance to tourism activity are discussed in chapters 6 and 7.
55. Campelo, 2017. Page 7.
56. Law, 2002.
57. MacLaran, A. (ed.) (2003) *Making Space: Property Development and Urban Planning*. London: Routledge.
58. MacLaran, 2003.
59. MacLaran, 2003.
60. Hall, T., & Barrett, H. (2018) *Urban Geography* (fifth edition). London: Routledge.
61. Hall & Barrett, 2018.
62. Smith, A. (2012) *Events and Urban Regeneration: The Strategic Use of Events to Revitalise Cities*. London: Routledge.
63. Tallon, 2013.
64. Hall & Barrett, 2018.
65. Roberts, P. (2000) The evolution, definition and purpose of urban regeneration. In P. Roberts and H. Sykes (eds.), *Urban Regeneration* (pp. 9–36). London: Sage. Page 17.
66. Tallon, 2013.
67. Tallon, 2013.
68. Hall & Barrett, 2018.
69. Gospodini, A. (2009) Post-industrial trajectories of mediterranean European cities: The case of post-Olympics Athens. *Urban Studies* 46(5&6), 1157–1186.
70. Gospodini, 2009.
71. Cadell, C., Falk, F., & King, F. (2008) *Regeneration in European Cities: Making Connections*. York: Joseph Rowntree Foundation.
72. Burgen, S. (2020, October 20) La Rambla: plans to transform Barcelona's tourist rat run into a cultural hub. *Guardian*. Available at: https://www.theguardian.com/travel/2020/oct/20/la-rambla-transform-barcelona-tourist-rat-run-into-a-cultural-hub (accessed 12 December 2020).
73. 'Case Study CC3: The regeneration of Lisbon's city centre' is sourced from Santos (2019) and Richards & Marques (2018).
74. Santos, J. R. (2019). Public space, tourism and mobility: Projects, impacts and tensions in Lisbon's urban regeneration dynamics. *Journal of Public Space* 4(2), 29–56, DOI: 10.32891/jps.v4i2.1203.
75. Santos, 2019.
76. The three network of public spaces identified by Santos (2019) are: (i) the riverfront system: Santos – Cais do Sodré – Ribeira das Naus – Terreiro do Paço

– Campo das Cebolas – Santa Apolónia; (ii) the garden and belvedere system: Príncipe Real – São Pedro de Alcântara – Portas do Sol – Graça – Senhora do Monte; (iii) the street and square system: Santa Catarina – Bica – Chiado – Baixa- Rossio – Martim Moniz – Mouraria – Castelo – Alfama.

77. MAAT – Museum of Art, Architecture and Technology. Architects: Amanda Levete Architects.

78. Richards & Marques, 2018.

79. 'Case Study CC2: Regenerating Sheffield' is sourced from: Booth, P. (2010) Sheffield: a miserable disappointment no more? In J. Punter (ed.), *Urban Design and the British Urban Renaissance* (pp. 85–99). London: Routledge.

80. Vincent, J. (2014, March 21) Sheffield Supertram marks twenty-year journey. *BBC Look North*. Available at https://www.bbc.com/news/uk-england-south-yorkshire-26663137 (accessed 28 February 2021).

# CHAPTER 3

1. Dunne, G., Flanagan, S., & Buckley, J. (2010) Towards an understanding of International City Break Travel. *International Journal Tourism Research* 12, 409–417.

2. Dunne et al., 2010.

3. Trew, J., & Cockerell, N. (2002) The European market for UK city breaks. *Insights* 14(58), 85–111. Page 86.

4. Sacco, P. L. (2011) *Culture 3.0: A New Perspective for the EU 2014–2020 Structural Funds Programming*. Report for the OMC Working Group on Cultural and Creative Industries.

5. Richards, G., & Marques, L. (2018) *Creating Synergies between Cultural Policy and Tourism for Permanent and Temporary Citizens*. Available at: http://www.agenda21culture.net (accessed 15 February 2021).

6. Sacco, 2011.

7. Richards, G. (2014) Cultural tourism 3.0. The future of urban tourism in Europe? In R. Garibaldi (ed.), *Il turismo culturale europeo. Città ri-visitate. Nuove idee e forme del turismo culturale* (pp. 25–38). Milan: Franco Angeli.

8. The significant investments in museums in the 1990s and 2000s is discussed in section 10.2.

9. Richards & Marques, 2018.

10. Nientied, P. (2020) Rotterdam and the question of new urban tourism. *International Journal of Tourism Cities*. DOI: 10.1108/IJTC-03-2020-0033.

11. Crespi-Vallbona, M., & Pérez, M. D. (2015) Tourism and food markets: A typology of food markets from case studies of Barcelona and Madrid. *Regions Magazine* 299(1), 15–17.

12. Crespi-Vallbona & Pérez, 2015.

13. Richards, G. (2017) Tourists in their own city: Considering the growth of a phenomenon. *Tourism Today* 16, 8–16.

14. Maitland, R., & Newman, P. (2014) Developing world tourism cities. In R. Maitland & P. Newman (eds.), *World Tourism Cities: Developing Tourism Off the Beaten Track* (pp. 1–21). London: Routledge.

15. Maitland, R. (2008) Conviviality and everyday life: the appeal of new areas of London for visitors. *International Journal of Tourism Research* 10, 15–25.

16. Richards, 2014.

17. Richards, G., & Marques, L. (2018) *Creating Synergies between Cultural Policy and Tourism for Permanent and Temporary Citizens.* Available at: http://www.agenda21culture.net (accessed on 15 February 2021).

18. Short-term rental tourism accommodation is discussed in greater detail in chapter 6.

19. Richards, 2014.

20. Law, 2002.

21. Campisi, D., Costa, R., & Mancuso, P. (2010) The effects of low-cost airlines growth in Italy. *Modern Economy* 1, 59–67

22. Assaeroporti (n.d.) Dati Annuali. Retrieved from https://assaeroporti.com/dati-annuali/.

23. Ciesluk, K. (2020, June 14) *How Do Low Cost Carriers Actually Make Money: A Complete Breakdown.* Retrieved from https://simpleflying.com/how-low-cost-carriers-make-money/.

24. Ebejer, J., Xuereb, K., & Avellino, M. (2020) A critical debate of the cultural and social effects of Valletta 2018 European Capital of Culture. *Journal of Tourism and Cultural Change,* DOI: 10.1080/14766825.2020.1849240.

25. European Commission (n.d.) Transport and mobility: Trans-European Transport Network (TEN-T). Retrieved from: https://ec.europa.eu/transport/themes/infrastructure/ten-t_en.

26. 'Case Study CC1 The total reconstruction Rotterdam Centraal station' is sourced from: Griffiths, A. (2014, March 22) Rotterdam Centraal station reopens with a pointed metal-clad entrance. *DeZeen.* Available at: https://www.dezeen.com/2014/03/22/rotterdam-centraal-station-benthem-crouwel-mvsa-architects-west-8/ (accessed 18 February 2021).

27. Three architectural firms were jointly responsible for designing Rotterdam Centraal and its surroundings: Benthem Crouwel Architects, MVSA Meyer en van Schooten Architecten, West 8.

28. Cadell, C., Falk, F., & King, F. (2008) *Regeneration in European Cities: Making Connections.* York: Joseph Rowntree Foundation.

29. Lassen, C., Smink, C., & Smidt-Jensen, S. (2009) Experience spaces, (aero) mobilities and environmental impacts. *European Planning Studies* 17(6), 887–903.

30. Lassen et al., 2009.

31. The cities across Europe with direct flights to Billund up until 2019: Stockholm (Sweden); Bergen, Oslo, Stavanger (Norway); Helsinki (Finland); Reykjavik (Iceland); Brussels (Belgium); Amsterdam (The Netherlands); Paris (France); Dublin (Ireland); Edinburgh, London, Manchester (UK); Düsseldorf, Frankfurt, Berlin (Germany); Prague (Czech Republic); Vienna, Innsbruck (Austria); Vilnius (Lithuania); Riga (Latvia); Warsaw, Gdańsk, Kraków (Poland); Kiev, Lviv

(Ukraine); Bergamo, Pisa, Rome (Italy); Barcelona, Málaga (Spain); Tuzla (Bosnia and Herzegovina); Budapest (Hungary); Sofia, Varna, Burgas (Bulgaria); Bacău, Bucharest, (Romania); Istanbul, Antalya (Turkey); Larnaka (Cyprus); and Malta.

32. Wiki2 (n.d.) *List of the busiest airports in the Nordic countries.* Retrieved from https://wiki2.org/en/List_of_the_largest_airports_in_the_Nordic_countries#2019_statistics (accessed 15 January 2021).

33. TEA/AECOM (2020) *TEA/AECOM 2019 Theme Index and Museum Index: The Global Attractions Attendance Report.* Themed Entertainment Association. Available at: https://aecom.com/theme-index/ (accessed 2 February 2020).

## CHAPTER 4

1. UNWTO – World Tourism Organization & ETC – European Travel Commission (2011). *Handbook on Tourism Product Development.* Madrid: WTO.

2. Ashworth G. J., & Tunbridge, J. E. (2000) *The Tourist-Historic Cities: Retrospect and Prospect of Managing the Heritage City.* London: Routledge.

3. UNWTO, 2011. Page 2.

4. Kolb, B. (2017) *Tourism Marketing for Cities and Towns: Using Social Media and Branding to Attract Tourists.* London: Routledge.

5. Law, C. (2002) *Urban Tourism: The Visitor Economy and the Growth of Large Cities* (second edition). New York: Continuum.

6. Selby, M. (2004) *Understanding Urban Tourism: Image, Culture and Experience.* London: I. B. Tauris.

7. Kolb, 2017.

8. Page, S. (2011) *Tourism Management: An Introduction* (fourth edition). Oxford, UK, & Burlington, MA: Butterworth-Heinemann.

9. Page, 2011.

10. Maitland, R. (2006) How can we manage the tourist-historic city? Tourism strategy in Cambridge, UK, 1978–2003. *Tourism Management* 27(6), 1262–1273.

11. Shaw, G., & Williams, A. (2002) *Critical Issues in Tourism: A Geographical Perspective* (second edition). Oxford, UK: Blackwell.

12. Ashworth & Tunbridge, 2000.

13. Ashworth & Tunbridge, 2000.

14. Ashworth & Tunbridge, 2000. Page 59.

15. Ritchie, J., & Crouch, G. (2003). *The Competitive Destination: A Sustainable Tourism Perspective.* Cambridge, MA: CABI Publishing.

16. Ritchie & Crouch, 2003.

17. UNWTO, 2011.

18. UNWTO, 2011.

19. UNWTO, 2011.

20. UNWTO, 2011.

21. Kolb, 2017.

22. Specht, J. (2014) *Architectural Tourism: Building for Urban Travel Destinations.* Wiesbaden: Springer Gabler.

23. Specht, 2014.

24. Suvantola, J. (2002) *Tourist's Experience of Place*. Aldershot: Ashgate.

25. Judd, D. (1999) Constructing the tourist bubble. In D. Judd & S. Fainstein (eds.), *The Tourist City* (pp. 35–53). New Haven, CT: Yale University Press.

26. Selby, 2004.

27. Hayllar, B., & Griffin, T. (2007) A tale of two precincts. In J. Tribe & D. Airey (eds.), *Developments in Tourism Research*. Oxford, UK: Elsevier. Page 155.

28. Shaw & Williams, 2002.

29. Ashworth, G., & Page, S. (2011) Urban tourism research: Recent progress and current paradoxes. *Tourism Management* 32, 1–15.

30. Maitland, R. (2008) Conviviality and everyday life: the appeal of new areas of London for visitors. *International Journal of Tourism Research* 10, 15–25.

31. Specht, 2014.

32. Woodward, R. (2005) Water in landscape. In H. Dreiseitl & D. Grau (eds.), *New Waterscapes; Planning, Building and Designing with Water*. Basel: Birkauser.

33. Curtis, S. (2019) The river Thames: London's Riparian highway. In A. Smith and A. Graham (eds.), *Destination London: The Expansion of the Visitor Economy* (pp. 165–182). London: University of Westminster Press.

34. Shydlovska, K. (2019) *Outdooractive: Multimedia Fountain*. Available at: https://www.outdooractive.com/en/poi/wroclaw/multimedia-fountain/41228551/ (accessed 8 November 2020).

35. Shydlovska, 2019.

36. The Miroir d'Eau was designed by landscape artist Michel Corajoud and was built by the fountain-maker Jean-Max Llorca and the architect Pierre Gangnet.

37. Canepa, S., & Ab Ghafar, N. (2020) Water in architecture, architecture of water. *Journal of Civil Engineering and Architecture* 14, 249–262.

38. Princess Diana Memorial Fountain, design architects: Gustafson Porter + Bowman.

39. Stevens, Q. (2009) Nothing more than feelings. *Architectural Theory Review* 14(2), 156–172.

40. Archdaily (2017) Diana, Princess of Wales Memorial Fountain. *ArchDaily*, 1 February. Available at: https://www.archdaily.com/803509/diana-princess-of-wales-memorial-fountain-gustafson-porter-plus-bowman (accessed 8 November 2020).

41. Hayllar & Griffin, 2007.

42. Ashworth G. J., & Tunbridge, J. E. (2000) *The Tourist-Historic Cities: Retrospect and Prospect of Managing the Heritage City*. London: Routledge. The first edition of the book was published in 1990.

43. Hayllar & Griffin, 2007.

44. Prideaux, B. (2009) *Resort Destinations. Evolution, Management and Development*. London: Routledge.

45. Richards, G., & Palmer, R. (2010) *Eventful Cities: Cultural Management and Urban Revitalisation*. London: Routledge.

46. Prideaux, 2009.

47. Spirou, C. (2011) *Urban Tourism and Urban Change: Cities in a Global Economy*. New York: Routledge.

48. Spirou, 2011. Page 80.
49. Spirou, 2011.
50. Spirou, 2011.
51. Hayllar & Griffin, 2007.
52. Hayllar & Griffin, 2007.
53. Also relevant to this section is the discussion on urban regeneration in chapter 2 and the discussion on the conservation of historic areas in chapter 9.
54. 'Case Study AR2: Liverpool's waterfront development' is sourced from: Spirou, C. (2011) *Urban Tourism and Urban Change: Cities in a Global Economy*. New York: Routledge.

## CHAPTER 5

1. Holloway, C. J., & Taylor, N. (2006) *The Business of Tourism* (seventh edition). Harlow: Financial Times & Prentice Hall.
2. Keyser, H. (2002) *Tourism Development*. Oxford, UK: Oxford University Press. Page 170.
3. Ebejer, J. (2015) *Tourist Experiences of Urban Historic Areas Valletta as a Case Study* (Doctoral thesis). University of Westminster.
4. Leask, A. (2008) The nature and role of visitor attractions. In A. Fyall, B. Garrod, A. Leask & S. Wanhill (eds.), *Managing Visitor Attractions: New Directions* (second edition). Oxford, UK: Butterworth-Heinemann.
5. Leiper, N. (2004). *Tourism Management* (third edition). Sydney: Pearson Education Australia.
6. Keyser, 2002.
7. Holloway & Taylor, 2006. Page 218.
8. Getz, D. (2016) *Event Studies: Theory, Research and Policy for Planned Events* (third edition). London: Routledge.
9. Lewis, G. D. (2020, November 6) Museum. *Encyclopædia Britannica*. Available at: https://www.britannica.com/topic/museum-cultural-institution (accessed on 1 January 2021).
10. Macleod, S., Hourston Hanks, L., & Hale, J. (2012) *Museum Making: Narrative, Architectures, Exhibitions*. London: Routledge. Page xix.
11. Barreneche, R. A. (2005) *New Museums*. London: Phaidon.
12. Section 9.3 discusses the difficulties involved in adapting historic buildings to modern-day use.
13. Cellini, R., & Cuccia, T. (2018) How free admittance affects charged visits to museums: an analysis of the Italian case. *Oxford Economic Papers* 70(3), 680–698.
14. BBC (2011) *Museums enjoy 10 years of freedom*. Available at: https://www.bbc.com/news/entertainment-arts-15927593. (Accessed 3 January 2021).
15. BBC, 2011.
16. BBC, 2011.
17. Cellini & Cuccia, 2018.
18. Cellini & Cuccia, 2018.

19. TEA/AECOM (2020) *TEA/AECOM 2019 Theme Index and Museum Index: The Global Attractions Attendance Report.* Themed Entertainment Association. Available at: https://aecom.com/theme-index/ (accessed 12 December 2020).

20. Information in figure 5.1 is sourced from: TEA/AECOM, 2020.

21. Staniewska, A. (2019) Translating fortified landscape heritage to the public: Dilemmas on rehabilitation, popularization and conservation methods. In R. Kusek & J. Purchla (eds.) *Heritage and Society* (pp. 189–204). Krakow: International Cultural Centre.

22. An urban area is said to be 'legible' when the street layout is such that it makes it easy for people who are not familiar to find their way with little risk of getting lost.

23. Ebejer, 2015.

24. See explanation of environmental legibility as per Lynch (1960) in chapter 8: Meaning of Place and the Tourist Experience.

25. Ashworth, G., & Bruce, D. (2009) Town walls, walled towns and tourism: Paradoxes and paradigms. *Journal of Heritage Tourism* 4(4), 299–313.

26. Ashworth & Bruce, 2009.

27. Pérez Guilarte, Y., & Rubén Camilo Lois González, R. (2018) Sustainability and visitor management in tourist historic cities: The case of Santiago de Compostela, Spain. *Journal of Heritage Tourism* 13(6), 489–505.

28. Jackson, S. (2018) London Mithraeum: Reimagining the famous Roman temple. *Current Archaeology* 334, December 1.

29. Holloway & Taylor, 2006.

30. Dick, M., Eveleigh, D., & Sullivan, J. (2019) *The Black Country: A History in 100 Objects.* West Midlands: Black Country Living Museum Publications.

31. Holloway & Taylor, 2006.

32. Holloway & Taylor, 2006.

33. 'Case Study VA2: The Guinness Storehouse in Dublin' is sourced from: Müller, S., & Brunner-Sperdin, A. (2006) Entertainment, attraction and engagement: Pre-condition for the new holiday experience and leisure time! In K. Weiermair, P. Pechlaner, & T. Bieger (eds.), *Time Shift, Leisure and Tourism: Impacts of Time Allocation on Successful Products and Services* (pp. 249–265). Berlin: Erich Schmidt Verlag GmbH & Co.

34. Müller & Brunner-Sperdin, 2006.

35. Pine, B., & Gilmore, J. (1999) *The Experience Economy.* Boston: Harvard Business School Press.

36. Kolb, B. (2017) *Tourism Marketing for Cities and Towns: Using Social Media and Branding to Attract Tourists.* London: Routledge.

37. Ebejer, 2015.

38. Dimitrovskia D., & Crespi-Vallbona, M. (2018) Urban food markets in the context of a tourist attraction: La Boqueria market in Barcelona. *Tourism Geographies* 20(3), 397–417.

39. Crespi-Vallbona, M., & Pérez, M. D. (2015) Tourism and food markets: A typology of food markets from case studies of Barcelona and Madrid. *Regions Magazine* 299(1), 15–17.

40. Gilli, M., & Ferrari, S. (2018) Tourism in multi-ethnic districts: The case of Porta Palazzo market in Torino. *Leisure Studies* 37(2), 146–157.

41. Gilli & Ferrari, 2018.

42. Punter, D. (2019, 12 February) Ulysses. *Encyclopedia Britannica*. Available at: https://www.britannica.com/topic/Ulysses-novel-by-Joyce (accessed 21 March 2021).

43. Kolb, 2017.

44. The history of Europe is closely associated with the history of Christianity; the Catholic faith in particular. This is why the majority of religious buildings in Europe are of the Catholic tradition. It is for this reason that this section mostly makes reference to Catholic churches and cathedrals.

45. Woodward, S. (2004) Faith and tourism: planning tourism in relation to places of worship. *Tourism and Hospitality Planning and Development* 1(2), 173–186.

46. Notre-Dame Cathedral was severely damaged in a fire that broke out on 15 April 2019. Most of the roof, a spire and some of the rib vaulting were destroyed.

47. Smørvik, K. (2020) Why enter the church on holiday? Tourist encounters with the Basilica of Santa Maria in Trastevere, Rome. *Journal of Heritage Tourism*. DOI: 10.1080/1743873X.2020.1807557.

48. Lupu, C., Brochado, A., & Stoleriu, O. (2019). Visitor experiences at UNESCO monasteries in Northeast Romania. *Journal of Heritage Tourism* 14(2), 150–165.

49. Hughes, K., Bond, N., & Ballantyne, R. (2013). Designing and managing interpretive experiences at religious sites: Visitors' perceptions of Canterbury Cathedral. *Tourism Management* 36, 210–220.

50. Lupu et al., 2019.

51. Winter, M., & Gasson, R. (1996). Pilgrimage and tourism: cathedral visiting in contemporary England. *International Journal of Heritage Studies* 2(3), 172–182.

52. Smørvik, 2020.

53. Smørvik, 2020. Page 3.

54. Smørvik, 2020.

55. Hughes et al., 2013. Page 218.

56. Smørvik, 2020. Page 8.

57. Lupu et al., 2019.

58. Oriade, A., & Cameron, H. (2018) Cathedral Tourism. In S. Agarwal, G. Busby & R. Huang (eds.) *Special Interest Tourism: Concepts, Contexts and Cases* (pp. 40–51). Wallingford: CABI.

59. Oriade & Cameron, 2018.

60. Oriade & Cameron, 2018.

61. Uggla, Y. (2014) Protecting urban greenery: The case of Stockholm's National City Park. *City and Community,* december, 13(4).

62. Connell, J., & Meyer, D. (2004) Modelling the visitor experience in the gardens of Great Britain. *Current Issues in Tourism* 7(3), 183–216.

63. Connell & Meyer, 2004.

64. Britannica (n.d.) *Botanical Garden.* Available at: https://www.britannica.com/science/botanical-garden-study-and-exhibition-garden (accessed 15 January 2021).

65. Benfield, R. (2018) Garden tourism. In S. Agarwal, G. Busby, & R. Huang (eds.), *Special Interest Tourism: Concepts, Contexts and Cases* (pp. 156–170). Wallingford: CABI.

66. VisitBritain (n.d.) Official Statistics. Available at: https://www.visitbritain.org/official-statistics.

67. Smith, A. (2019) Event takeover? The commercialisation of London's Parks. In A. Smith & A. Graham (eds.), *Destination London: The Expansion of the Visitor Economy* (pp. 205–224). London: University of Westminster Press.

68. 'Case Study VA4' is sourced from: Weddle Landscape Design (n.d.) Sheffield Winter Gardens. Available at: http://weddles.co.uk/portfolio/sheffield-winter-gardens/ (accessed 16 January 2021).

69. 'Case Study VA5: Stockholm's National City Park' is sourced from: Uggla, Y. (2014) Protecting urban Greenery: the case of Stockholm's National City Park. *City & Community*, December, 13(4).

70. Uggla, Y., & Olausson, U. (2013) The enrollment of nature in tourist information: framing urban nature as 'the other'. *Environmental Communication* 7(1), 97–112.

71. Mason, P. (2000) Zoo tourism: The need for more research. *Journal of Sustainable Tourism* 8(4), 333–339.

72. Shackley, M. (1996) *Wildlife Tourism*. London: Routledge.

73. A butterfly park is a greenhouse that provides a favourable atmosphere for butterflies to grow. These are often open to the public as visitor attractions.

74. Mason, 2000.

75. World Cities Ranking (n.d.) *Largest and Best Aquariums in Europe 2021.* Available at https://worldcitiesranking.com/largest-and-best-aquariums-in-europe-top-10/ (accessed on 26 February 2021).

76. Holloway & Taylor, 2006.

77. Holloway & Taylor, 2006.

78. TEA/AECOM, 2020.

79. Other parks worthy of note are Tivoli Gardens and Legoland Billund in Denmark; Portaventura in Spain; Liseberg in Sweden; Gardaland in Italy; Parc Asterix (near Paris), and Futuroscope (near Poitiers) in France; Alton Towers, Thorpe Park, Chessington World of Adventures and Legoland Windsor in the UK; Phantasialand, Heidi Park and Legoland Deutschland in Germany. For each, the number of visitors is between 1.5 million and 3 million per year.

80. TEA/AECOM, 2020.

81. Holloway & Taylor, 2006.

# CHAPTER 6

1. Ashworth G. J., & Tunbridge, J. E. (2000) *The Tourist-Historic Cities: Retrospect and Prospect of Managing the Heritage City*. London: Routledge.

2. Fletcher, J., Fyall, A., Gilbert, D., & Wanhill, S. (eds.) (2013) *Tourism: Principles and Practice* (fifth edition). Pearson.

3. Page, S. (2011) *Tourism Management: An Introduction* (fourth edition). Oxford, UK: Elsevier.

4. Fletcher et al., 2013.
5. Rutes, W., Penner, R., & Adams, L. (2001) *Hotel Design: Planning and Development*. New York: W. W. Norton & Company.
6. D'Aveni, R. (2010) How to escape the differentiation proliferation trap. *Strategy and Leadership* 38(3), 44–49.
7. Penner, R., Adams, L. & Robinson, S. (2013), *Hotel Design Planning and Development* (second edition). New York: W. W. Norton & Company.
8. Law, C. (2002) *Urban Tourism: The Visitor Economy and the Growth of Large Cities* (second edition). New York: Continuum.
9. Page, 2011.
10. Richards, G. (2016) Hostels and the making of urban spaces. In P. Russo and G. Richards (eds.), *Reinventing the Local in Tourism: Producing, Consuming and Negotiating Place* (pp. 171–184). Bristol: Channel View Publications.
11. Gilli, M., & Ferrari, S. (2016), The 'Diffuse Hotel': An Italian new model of sustainable hospitality. In A. Russo & G. Richards (eds.), *Reinventing the Local in Tourism: Producing, Consuming and Negotiating Place* (pp. 65–83). Bristol: Channel View Publications.
12. Penner et al., 2013.
13. Correira Loureiro, S., Rita, P., & Moraes Sarmento, E. (2020) What is the core essence of small city boutique hotels? *International Journal of Culture, Tourism and Hospitality Research* 14(1), 44–62.
14. Anhar, L. (2001) *The definition of boutique hotels*. Available at: www.hospitalitynet.org/news/4010409.html (accessed 18 June 2018).
15. Lim, W. M., & Endean, M. (2009) Elucidating the aesthetic and operational characteristics of UK boutique hotels. *International Journal of Contemporary Hospitality Management* 21(1), 38–51.
16. Jones, D., Day, J., & Quadri-Felitti, D. (2013) Emerging definitions of boutique and lifestyle hotels: A Delphi study. *Journal of Travel and Tourism Marketing* 30(7), 715–731.
17. Correira Loureiro et al., 2020.
18. Jones et al., 2013.
19. Fletcher et al., 2013
20. Penner et al., 2013.
21. Pasquinelli, C. (2017) Tourism connectivity and spatial complexity: A widening bi-dimensional arena of urban tourism research. In N. Bellini & C. Pasquinelli (eds.), *Tourism in the City. Towards an Integrative Agenda on Urban Tourism* (pp. 29–50). Switzerland: Springer.
22. Füller, H., & Michel B. (2014) 'Stop Being a Tourist!' New dynamics of urban tourism in Berlin Kreuzberg. *International Journal of Urban and Regional Research* 38, 1304–1318.
23. Guttentag, D. (2015) Airbnb: Disruptive innovation and the rise of an informal tourism accommodation sector. *Current Issues in Tourism* 18(12), 1192–1217.
24. Guttentag, 2015.
25. Calle-Vaquero, M.d.l., García-Hernández, M., & Mendoza de Miguel, S. (2021) Urban planning regulations for tourism in the context of overtourism: Applications in historic centres. *Sustainability* 13, 70.

26. Füller & Michel, 2014.
27. Law, 2002.
28. Ashworth & Tunbridge, 2000.
29. Law, 2002.
30. Timothy, D. J. (2005) *Shopping, Tourism, Retail and Leisure*. Clevedon: Channel View Publications.
31. Goeldner, C. R., & Ritchie, J. R. B. (2009). *Tourism: Principles, Practices, Philosophies*. Hoboken, NJ: Wiley.
32. Rabbiosi, C. (2015) Renewing a historical legacy: Tourism, leisure shopping and urban branding in Paris. *Cities* 42, 195–203.
33. Westwood, S. (2006) Shopping in sanitised and un-sanitised spaces. Adding value to the tourist experiences. *Journal of Retail and Leisure Property* 5(4): 281–291.
34. Timothy, D. J. (2014) Trends in tourism, shopping and retailing. In A. Lew, M. Hall & A. Williams (eds.), *The Wiley Blackwell Companion to Tourism* (pp. 378–388). Oxfore, UK: Wiley Blackwell.
35. Timothy, 2014.
36. Westwood, 2006.
37. Orbasli, A. (2000) *Tourists in Historic Towns: Urban Conservation and Heritage Management*. London: E&FN Spon.
38. Specht, J. (2014) *Architectural Tourism: Building for Urban Travel Destinations*. Wiesbaden: Springer Gabler.
39. Orbasli, 2000.
40. Law, 2002.
41. Orbasli, 2000.
42. In the text, the term 'conferences' includes 'congresses' and 'conventions,' which is the terminology used by some business travel publications. In some texts, smaller conferences are referred to as 'meetings'.
43. Davidson, R., & Cope, B. (2003) *Business Travel: Conference, Incentive Travel, Exhibitions, Corporate Hospitality and Corporate Travel*. Harlow: Pearson Education Limited.
44. Rogers, T. (1998) *Conferences: A Twenty-First Century Industry*. Harlow: Addison Wesley Longman.
45. Davidson & Cope, 2003.
46. Rogers, 1998.
47. Davidson & Cope, 2003.
48. Davidson & Cope, 2003.
49. Davidson & Cope, 2003.

# CHAPTER 7

1. Pavia, A. (2018) Walking as a luxury activity. In M. Paris (ed.), *Making Prestigious Places: How Luxury Influences the Transformation of Cities* (pp. 73–80). London: Routledge.

2. Pavia (2008). Page 73.
3. O'Mara, S. (2019) *In Praise of Walking: The New Science of How We Walk and Why It's Good for Us*. London: The Bodley Head. Page 122.
4. Landry, C. (2006) *The Art of City Making*. London: Earthscan. Page 176.
5. Wunderlich, F. M. (2008) Walking and rhythmicity: Sensing urban space. *Journal of Urban Design* 13(1), 125–139. Page 128.
6. Specht, J. (2014) *Architectural Tourism: Building for Urban Travel Destinations*. Wiesbaden: Springer Gabler.
7. Burns, A. (2000) Emotion and urban experience: Implications for design. *Design Issues* 16(3), 67–79.
8. Orbasli, A. (2000) *Tourists in Historic Towns: Urban Conservation and Heritage Management*. London: E&FN Spon.
9. Appleyard, D. (1979) *The Conservation of European Cities*. Cambridge, MA: MIT Press. Page 19.
10. Orbasli, 2000.
11. Wunderlich (2008).
12. Wunderlich (2008). Page 132.
13. Sinha, A., & Sharma, Y. (2009) Urban design as a frame for site readings of heritage landscapes: A case study of Champaner-Pavagadh, Gujarat, India. *Journal of Urban Design* 14(2), 203–221.
14. Staiff, R. (2012) The somantic and the aesthetic: Embodied heritage tourism experiences of Luang Prabang, Lao. In L. Smith, E. Waterton & S. Watson (eds.), *The Cultural Moment in Tourism*. London: Routledge. Page 45.
15. Stevenson, N., & Farrell, H. (2017) Taking a hike: exploring leisure walkers embodied experiences. *Social and Cultural Geography* 19(4), 429–447.
16. Ebejer, J. (2015) *Tourist Experiences of Urban Historic Areas: Valletta as a Case Study* (Doctoral thesis). University of Westminster.
17. Wunderlich (2008).
18. Ross, G. F. (1994). *The Psychology of Tourism*. Melbourne: Hospitality Press.
19. Ross (1994).
20. Ebejer, J. (2015).
21. Suvantola, J. (2002) *Tourist's Experience of Place*. Aldershot: Ashgate. Page 180.
22. Hayes, D., & Macleod, N. (2007). Packaging places: Designing heritage trails using an experience economy perspective to maximize visitor engagement. *Journal of Vacation Marketing* 13(1), 45–58. Page 48.
23. O'Mara, 2019. Page 99.
24. Southworth, M. (2005) Designing the walkable city. *Journal of Urban Planning and Development* 4(1), 246–257.
25. O'Mara, 2019.
26. Hall, C. M., & Ram, Y. (2019) Measuring the relationship between tourism and walkability? Walk Score and English tourist attractions. *Journal of Sustainable Tourism* 27(2), 223–240.
27. Southworth, 2005. Page 248.

28. Henderson, J. (2018) Making cities more walkable for tourists: A view from Singapore's streets. *International Journal of Tourism Cities* 4(3).
29. Southworth, 2005.
30. Southworth, 2005.
31. Southworth, 2005.
32. O'Mara, 2019. Page 102.
33. Hall & Ram (2019).
34. Curtis, S. (2019) The river Thames: London's Riparian highway. In A. Smith and A. Graham (eds.), *Destination London: The Expansion of the Visitor Economy* (pp. 165–182). London: University of Westminster Press.
35. Millennium Bridge: Architects: Foster & Partners. Engineers: Ove Arup.
36. *Guardian* (2000, 12 June) *Swaying Millennium Bridge to close after two days*. Available at: https://www.theguardian.com/uk/2000/jun/12/2 (accessed 12 February 2021).
37. Maitland, R., & Newman, P. (2009) London: Tourism moving east? In R. Maitland & P. Newman (eds.), *World Tourism Cities: Developing Tourism off the Beaten Track* (pp. 66–86). London: Routledge.
38. Maitland & Newman (2009).
39. Samarasekara, G. N., Fukahori, K., & Kubota, Y. (2011) Environmental correlates that provide walkability cues for tourists: An analysis based on walking decision narrations. *Environment and Behaviour* 43(4), 501–524.
40. Timothy, D. J., & Boyd, S. W. (2015) *Tourism and Trails: Cultural, Ecological and Management Issues*. Bristol: Channel View.
41. Ashworth, G., & Bruce, D. (2009) Town walls, walled towns and tourism: Paradoxes and paradigms. *Journal of Heritage Tourism* 4(4), 299–313.
42. Bruce, D., & Creighton, O. (2006) Contested identities: The dissonant heritage of European town walls and walled towns. *International Journal of Heritage Studies* 12(3), May, 234–254. Page 240.
43. Ashworth & Bruce, 2009. Page 303.
44. Timothy & Boyd (2015).
45. Hass-Klau, C. (1990) *The Pedestrian and City Traffic*. London: Belhaven Press.
46. Hass-Klau, 1990.
47. Hass-Klau, 1990.
48. Hass-Klau, 1990.
49. Hass-Klau, 1990.
50. Hass-Klau, 1990.
51. Rodriguez, L. (2011, January 3) Pedestrian-only shopping streets make communities more livable. *Planetizen*. Available at: https://www.planetizen.com/node/47517 (accessed 20 February 2021).
52. Bergenhenegouwen, G., & van Weesep, J. (2003) Manipulated space: The Beurstraverse retail complex in Rotterdam. *Belgeo* 1, 79–86.
53. The actual name of the retail complex is Beurstraverse.
54. Suvantola, 2002.

55. Garrod B (2009) Understanding the relationship between tourism destination imagery and tourist photography. *Journal of Travel Research* 47(3), 346–358.
56. Prideaux, B., & Coghlan, A. (2010) Digital cameras and photo-taking behaviour on the Great Barrier Reef: Marketing opportunities for Reef tour operators. *Journal of Vacation Marketing* 16, 171–183.
57. Lee, D. H. (2010) Digital cameras, personal photography and the reconfiguration of spatial experiences. *Information Society* 26(4), 266–275. Page 270.
58. Van Dijck, J. (2008) Digital photography: Communication, identity, memory. *Visual Communication* 7(1), 57–76. Page 60.
59. Lee, 2010.
60. Lee, 2010. Page 272.
61. Andersson Cederholm, E. (2004) The use of photo-elicitation in tourism research: Framing the backpacker experience. *Scandinavian Journal of Hospitality and Tourism* 4(3), 225–241.
62. Andersson Cederholm, 2004. Page 236.
63. Van Dijck, 2008.
64. Crang, M. (1997) Picturing practices: research through the tourist gaze. *Progress in Human Geography* 21(3), 359–373. Page 363.
65. Andersson Cederholm, 2004. Page 239.
66. Suvantola, 2002.
67. Lee, 2010. Page 271.
68. Haldrup, M., & Larsen, J. (2003). The family gaze. *Tourist Studies* 3(1), 23–45.
69. Andersson Cederholm, 2004. Page 226.
70. Prideaux & Coghlan, 2010.
71. Lee, 2010.

# CHAPTER 8

1. Montgomery, J. R. (1998) Making a city: Urbanity, vitality and urban design. *Journal of Urban Design* 3(1), 93–116. Page 94.
2. The bulk of this chapter is sourced from Ebejer, J. (2015) *Tourist Experiences of Urban Historic Areas: Valletta as a Case Study* (Doctoral thesis). University of Westminster.
3. Berleant, A. (2003) The aesthetic in place. In S. Menin (ed.), *Constructing Place: Mind and Matter* (pp. 41–54). London: Routledge. Page 51.
4. Norberg-Schulz, C. (1980) *Genius Loci: Towards a Phenomenology of Architecture*. New York: Rizzoli.
5. Jiven, J., & Larkham, P. (2003) Sense of Place, authenticity and character. *Journal of Urban Design* 8(1), 67–82.
6. Hall, T., & Barrett, H. (2018) *Urban Geography* (fifth edition). London: Routledge.

7. Tuan, Y. F. (1977) *Space and Place: The Perspective of Experience.* Minneapolis: University of Minnesota Press.

8. Carmona, M., Heath, T., Oc, T., & Tiesdall, S. (2010) *Public Places: Urban Spaces: The Dimensions of Urban Design* (second edition). Oxford, UK: Architectural Press.

9. Hiss, T. (1991) *The Experience of Place.* New York: Knopf.

10. Carmona et al., 2010. Page 111.

11. Hall & Barrett, 2018.

12. Hall & Barrett, 2018.

13. Lynch, K. (1960) *The Image of the City.* Cambridge: MIT Press.

14. Lynch, 1960. Page 2.

15. Hall & Barrett, 2018.

16. Carmona et al., 2010.

17. Carmona et al., 2010.

18. Lynch, 1960.

19. Sinha, A., & Sharma, Y. (2009) Urban design as a frame for site readings of heritage landscapes: A case study of Champaner-Pavagadh, Gujarat, India. *Journal of Urban Design* 14(2), 203–221.

20. Montgomery, 1998.

21. Carmona et al., 2010.

22. Relph, 1976.

23. Castello, L. (2010) *Rethinking the Meaning of Place: Conceiving Place in Architecture-Urbanism.* Farnham: Ashgate.

24. Smaldone, D., Harris, C., & Sanyal, N. (2005) An exploration of place as a process: The case of Jackson Hole, WY. *Journal of Environmental Psychology* 25(4), December, 397–414. Page 400.

25. Creswell, J. W. (2007) *Qualitative Inquiry and Research Design: Choosing among Five Traditions* (second edition). Thousand Oaks, CA: SAGE.

26. Stokowski, P. (2002) Languages of place and discourses of power: Constructing new senses of place. *Journal of Leisure Research* 34(4), 368–382.

27. Stokowski, 2002. Page 369.

28. Griffin, T., & Hayllar, B. (2009) Urban tourism precincts and the experience of place. *Journal of Hospitality Marketing and Management* 18(2&3), 127–153. Page 147.

29. Uggla, Y., & Olausson, U. (2013) The enrollment of nature in tourist information: Framing urban nature as 'the other'. *Environmental Communication* 7(1), 97–112.

30. Canter, D. (1977) *The Psychology of Place.* London: Architectural Press.

31. Punter, J. (1991). Participation in the design of urban space. *Landscape Design* 200, 24–27.

32. Montgomery, 1998.

33. Gustafson, P. (2001) Meanings of place: Everyday experience and theoretical conceptualizations. *Journal of Environmental Psychology* 21(1), 5–16.

34. Carmona et al., 2010.

35. Relph, E. (1976) *Place and Placelessness.* London: Pion Limited.

36. Ebejer, J., Smith, A., Stevenson, N., & Maitland, R. (2020) The tourist experience of heritage urban spaces: Valletta as a case study. *Tourism Planning and Development* 17(4), 458–474.

37. Loeffler, T. (2004) A Photo elicitation study of the meanings of outdoor adventure experiences. *Journal of Leisure Research* 36(4), 536–556.

38. Ebejer et al., 2020.

39. Carmona et al., 2010.

40. Griffin, T., & Hayllar, B. (2007) Historic waterfronts as tourism precincts: an experiential perspective. *Tourism and Hospitality Research* 7(1), 3–16. Page 10.

41. Norberg-Schulz, C. (1988) *Architecture – Meaning and Place: Selected Essays*. New York: Rizzoli.

42. Relph, 1976.

43. Smaldone et al., 2005.

44. Berleant, 2003. Page 48.

45. Montgomery, 1998.

46. Montgomery, 1998. Page 98.

47. Griffin & Hayllar, 2007.

48. Eco, U. (1986) Function and sign: semiotics of architecture. In M. Gottdiener & A. Lagopoulos (eds.), *The City and the Sign*. Guildford, NY: Columbia University Press.

49. Cuthbert, A. (2006) *The Form of Cities*. Oxford, UK: Blackwell.

50. Mazumdar, S., & Mazumdar, S. (2004) Religion and place attachment: A study of sacred places. *Journal of Environmental Psychology* 24(3), September, 385–397.

51. Eco, 1986.

52. Cuthbert, 2006.

53. Strike, J. (1994) *Architecture in Conservation: Managing Developments at Historic Sites*. Oxford, UK: Routledge.

54. Grauman, C. F. (2002) The Phenomenological Approach to People-Environment Studies. In R. Bechtel & A. Churchman (eds.), *Handbook of Environmental Psychology*. New York: Wiley.

55. Berleant, 2003.

56. Staiff, R. (2012) The somantic and the aesthetic: Embodied heritage tourism experinces of Luang Prabang, Lao. In L. Smith, E. Waterton & S. Watson (eds.), *The Cultural Moment in Tourism*. London: Routledge.

57. Sancar, F., & Severcan, Y. (2010) Children's places: rural-urban comparisons using participatory photography in the Bodrum Peninsula, Turkey. *Journal of Urban Design* 15(3), 293–324.

58. Berleant, 2003.

59. Gospodini, A. (2001) Urban design, urban space morphology, urban tourism: An emerging new paradigm concerning their relationship. *European Planning Studies* 9(7), 925–934.

60. Wearing, B., & Wearing, S. L. (1996) Refocusing the tourist experience: The flaneur and the chorister. *Leisure Studies* 1, 229–243.

61. Tweed, C., & Sutherland, M. (2007) Built cultural heritage and sustainable urban development. *Landscape and Urban Planning* 83(1), 62–69.

62. Tunbridge, J., & Ashworth, G. (1996) *Dissonant Heritage: Managing the Past as a Resource in Conflict.* Chichester: John Wiley.

63. Staiff, 2012.

64. Voase, R. (2008) Rediscovering the imagination: Meeting the needs of the 'new' visitor. In A. Fyall, B. Garrod, A. Leask, & S. Wanhill (eds.), *Managing Visitor Attractions: New Directions* (second edition). Oxford, UK: Butterworth-Heinemann.

65. McIntosh, A. J. (1999) Into the tourist's mind: Understanding the value of the heritage experience. *Journal of Travel and Tourism Marketing* 8(1), 41–64. Page 57.

66. Sancar & Severcan, 2010.

67. Timothy, D. (1997) Tourism and the personal heritage experience. *Annals of Tourism Research* 24(3), 751–753.

68. Frankl, V. (1992) *Man's Search for Meaning: An Introduction to Logotherapy.* London: Rider.

69. Ragheb, M. (1996) The search for meaning in leisure pursuits: Review conceptualisation and a need for a psychometric development. *Leisure Studies* 15, 245–258.

70. Hannabus, S. (1999) Postmodernism and the heritage experience. *Library Management* 20(5), 295–302. Page 299.

71. Wang, N. (1999). Rethinking authenticity in tourism experience. *Annals of Tourism Research* 26(2), 349–370. Page 352.

72. Wearing & Wearing, 1996.

73. Voase, 2008.

74. Cohen, E. (1979) A phenomenology of tourist experiences. *Sociology* 13, 179–201.

75. Lengkeek, J. (2001) Leisure experience and imagination: Rethinking Cohen's modes of tourism experience. *International Sociology* 16(2), 173–184.

76. McIntosh, 1999.

77. Rickly-Boyd, J. (2009) The tourist narrative. *Tourist Studies* 9(3), 259–280.

78. Jamal, T., & Hollinshead, K. (2001) Tourism and the forbidden zone: The underserved power of qualitative inquiry. *Tourism Management* 22, 63–82. Page 73.

79. Rickly-Boyd, 2009. Page 262.

80. Chronis, A. (2012) Tourists as story-builders: Narrative construction at a Heritage Museum. *Journal of Travel and Tourism Marketing* 29(5), 444–459.

81. Rickly-Boyd, 2009.

82. Chronis, 2012.

83. Ashworth, G. (1995) Managing the cultural tourist. In G. Ashworth & A. Doetvorst (eds.), *Tourism and Spatial Transformations.* Wallingford: CAB International.

84. Ashworth, 1995. Page 270.

85. Relph, 1976. Page 49.

86. Ebejer, J. (2015) *Tourist Experiences of Urban Historic Areas: Valletta as a Case Study* (Doctoral thesis). University of Westminster.

87. Ebejer, 2015.
88. Ebejer, 2015.
89. Ebejer, J., & Tunbridge, J. (2020) The role of urban heritage in Malta's tourism: Issues and initiatives. In G. Cassar & M. Avellino (eds.), *Tourism and the Maltese Islands: Observations, Reflections and Proposals* (pp. 149–172). Malta: University of Malta. Institute for Tourism, Travel and Culture and Kite Group.
90. Maitland, R., & Smith, A. (2009) Tourism and the aesthetics of the built environment. In J. Tribe (ed.), *Philosophical Issues in Tourism*. Bristol: Channel View Publications.
91. Norberg-Schultz, 1988.
92. Berleant, 2003.
93. Chronis, 2012.
94. Ashworth, G., & Tunbridge, J. (2004). *Malta Makeover: Prospects for the Realignment of Heritage, Tourism and Development*. URSI Report 304, Urban and Regional Studies Institute, University of Groningen.
95. Ebejer & Tunbridge, 2020.
96. Urry, J. (2002). *The Tourist Gaze* (second edition). London: Sage.
97. Ek, R., Larsen, J., Hornskov, S., & Mansfeldt, O. (2008) A dynamic framework of tourist experiences: Space-time and performances in the experience economy. *Scandinavian Journal of Hospitality and Tourism* 8(2), 122–140.
98. Rickly-Boyd, J., & Metro-Roland, M. (2010) Background to the fore: The prosaic in tourist places. *Annals of Tourism Research* 37(4), 1164–1180.
99. Ek et al., 2008. Page 126.
100. Edensor, T. (2000) Staging tourism: tourists as performers. *Annals of Tourism Research* 27(2), 322–344.
101. Ek et al., 2008.
102. Ek et al., 2008. Page 125.
103. Wearing, S. L., & Wearing, B. (2001) Conceptualizing the selves of tourism. *Leisure Studies* 20(2), 143–159.
104. Edensor, 2000.
105. Rickly-Boyd & Metro-Roland, 2010.
106. Edensor, 2000.
107. McKercher, B., & du Cros. (2002) *Cultural Tourism: Partnership between Tourism and Cultural Heritage Management*. Binghamton, NY: Haworth.
108. Edensor, 2000.
109. Griffin & Hayllar, 2007. Page 13.
110. Wearing & Wearing, 1996.
111. Orbasli, A. (2000) *Tourists in Historic Towns: Urban Conservation and Heritage Management*. London: E&FN Spon. Page 55.
112. Maitland, R. (2007) Culture, city users and the creation of new tourism areas in cities. In M. K. Smith (ed.), *Tourism, Culture and Regeneration*. Wallingford: CABI.
113. Ashworth, 1995.
114. Griffin & Hayllar, 2007. Page 9.
115. Burns, A. (2000). Emotion and urban experience: Implications for design. *Design Issues* 16(3), 67–79. Page 73.

116. Kyle, G. T., Mowen A. J., & Tarrant, M. (2004) Linking place preferences with place meaning: An examination of the relationship between place motivation and place attachment. *Journal of Environmental Psychology* 24(4), 439–454.

# CHAPTER 9

1. Orbasli, A. (2000) *Tourists in Historic Towns: Urban Conservation and Heritage Management.* London: E&FN Spon.
2. Ashworth G. J., & Tunbridge, J. E. (2000) *The Tourist-Historic Cities: Retrospect and Prospect of Managing the Heritage City.* London: Routledge. Page 10.
3. Millar, P. (1995) Heritage management for heritage tourism. In S. Medlik (ed.), *Managing Tourism* (second edition). Oxford, UK: Butterworth-Heinemann.
4. Graham, B., Ashworth, G., & Tunbridge, J. (2016) *A Geography of Heritage.* London: Routledge. Page 2.
5. Tunbridge, J. (2018) Thirty years on have we resolved the heritage questions? *Heritage Quarterly* 32–33, 131–141.
6. Tunbridge, 2018.
7. Vahtikari, T. (2017) *Valuing World Heritage Cities.* London: Routledge.
8. Kovářová, L. (2017) *UNESCO World Heritage Sites: Ways of Presenting and Interpreting the Past* (PhD Thesis). Charles University in Prague.
9. Vahtikari, 2017.
10. Pérez Guilarte, Y., & Lois González, R. C. (2018) Sustainability and visitor management in tourist historic cities: The case of Santiago de Compostela, Spain. *Journal of Heritage Tourism* 13(6), 489–505.
11. Ashworth & Tunbridge, 2000.
12. Selby, M. (2004) *Understanding Urban Tourism: Image, Culture & Experience.* London: I. B. Taurus.
13. Ashworth & Tunbridge, 2000.
14. Orbasli, 2000.
15. Russell, D. (2003) Selling Bradford: Tourism and northern image in the late twentieth century. *Contemporary British History* 17(2), 49–68.
16. Ebejer, J., & Tunbridge, J. (2020) The role of urban heritage in Malta's tourism: Issues and initiatives. In G. Cassar & M. Avellino (eds.), *Tourism and the Maltese Islands Observations, Reflections and Proposals* (pp. 149–172). Malta: University of Malta. Institute for Tourism, Travel and Culture and Kite Group.
17. Orbasli, 2000.
18. Pendlebury, J. (2002) Conservation and regeneration: Complementary or conflicting processes? The case of Grainger Town, Newcastle upon Tyne. *Planning Practice and Research* 17(2), 145–158.
19. Ebejer, J., & Dimelli, D. (2019) *Conservation Issues of Two Fortified Historic Towns and World Heritage Sites: Rhodes and Valletta.* Conference paper at 5th Heritage Forum of Central Europe, International Cultural Centre, Krakow Poland. 19–20 September.
20. Bandarin, F., & Van Oers, R. (2012) *The Historic Urban Landscape: Managing Heritage in an Urban Century.* Chichester: Wiley-Blackwell.

21. Ebejer, J. (2019) Using fortifications for tourism: can conservation objectives be reconciled with financial sustainability? In R. Kusek & J. Purchla (eds.), *Heritage and Society* (pp. 353–366). Krakow, Poland: International Cultural Centre.
22. Orbasli, 2000.
23. Ashworth & Tunbridge, 2000.
24. Pérez Guilarte & Lois González, 2018.
25. Selby, 2004.
26. Ashworth & Tunbridge, 2000.
27. Maitland, R. (2006) How can we manage the tourist-historic city? Tourism strategy in Cambridge, UK, 1978–2003. *Tourism Management* 27(6), 1262–1273.
28. Dimelli, D. (2019) Modern conservation principles and their application in Mediterranean Historic Centres: The case of Valletta. *Heritage* 2, 787–796.
29. Orbasli, 2000.
30. Bandarin & Van Oers, 2012.
31. Ebejer & Dimelli, 2019.
32. Ebejer & Dimelli, 2019.
33. Pendlebury, 2002.
34. Pendlebury, 2002.
35. Hall, T., & Barrett, H. (2018) *Urban Geography* (fifth edition). London: Routledge.
36. Ashworth & Tunbridge, 2000.
37. Ashworth & Tunbridge, 2000.
38. Ashworth & Tunbridge, 2000.
39. Ashworth & Tunbridge, 2000.
40. Orbasli, 2000.
41. Orbasli, 2000.
42. Manley, S., & Guise, R. (1998) Conservation in the built environment. In C. Greed & M. Roberts (eds.), *Introducing Urban Design: Interventions and Responses* (pp 64–86). London: Routledge.
43. Ashworth & Tunbridge, 2000.
44. Ashworth & Tunbridge, 2000.
45. Ashworth & Tunbridge, 2000.
46. Ebejer & Tunbridge, 2020.
47. Orbasli, 2000.
48. Ashworth & Tunbridge, 2000.
49. Ebejer, 2019.
50. Ebejer & Tunbridge, 2020.
51. 'Case Study UC1: Restoration and adaptive reuse of fortifications in Malta' is sourced from: Ebejer & Tunbridge, 2020.
52. Vahtikari, 2017.
53. UNESCO World Heritage Convention (n.d.) *World Heritage List*. Available from https://whc.unesco.org/en/list/ (accessed 15 November 2020).
54. Winter, T. (2013) Cultures of interpretation. In R. Staiff, R. Bushell, & S. Watson (eds.), *Heritage and Tourism: Place, Encounter, Engagement* (pp. 172–186). London: Routledge.

55. Vahtikari, 2017.
56. González Santa-Cruza, F., & López-Guzmán, T. (2017) Culture, tourism and World Heritage Sites. *Tourism Management Perspectives* 24, 111–116.
57. Vahtikari, T. (2003) Urban Interpretations of World Heritage: re-defining the city. In Marjaana Niemi & Ville Vuolanto (eds.), *Reclaiming the City: Innovation, Culture, Experience*. Helsinki: Studia Fennica Historica.
58. World Heritage Sites that are individual building or a small group of buildings are *not* included in appendix B.
59. The texts of the brief descriptions in appendix B are sourced from UNESCO's official website (http://whc.unesco.org/en/list/). For more information on any of the inscribed urban areas, refer to the UNESCO's website.
60. Kovářová, 2017.
61. Kovářová, 2017.
62. González Santa-Cruza & López-Guzmán, 2017.
63. 'Case Study UC2: Suomenlinna Helsinki: Conservation and adaptive reuse' is sourced from: Ebejer (2019); Kim (2019); Governing Body of Suomenlinna (2000, 2012 & 2015) and Öystilä & Grönroos (2017).

# CHAPTER 10

1. Hall, T., & Barrett, H. (2018) *Urban Geography* (fifth edition). London: Routledge.
2. Specht, J. (2014) *Architectural Tourism: Building for Urban Travel Destinations*. Wiesbaden: Springer Gabler.
3. Eco, U. (1986) Function and sign: Semiotics of architecture. In M. Gottdiener and A. Lagopoulos (eds.), *The City and the Sign*. Guildford, NY: Columbia University Press.
4. Specht, 2014. Page 117.
5. Specht, 2014.
6. Specht, 2014.
7. Specht, 2014. Page 62.
8. Specht, 2014.
9. Marques, B., & McIntosh, J. (2018) The spell of the visual and the experience of the sensory: Understanding icons in the built environment. *Charrette* 5(1) Spring.
10. Specht, 2014.
11. Specht, 2014.
12. Scerri, M., Edwards, D., & Foley, C. (2019) Design, architecture and the value to tourism. *Tourism Economics* 25(5), 695–710.
13. Leiper, N., & Park, S-Y. (2010) Skyscrapers' influence on cities' roles as tourist destinations. *Current Issues in Tourism* 13(4), 333–349. Page 333.
14. Smith, A. (2019) Vertical city tourism: Heightened aesthetic and kinaesthetic experiences. In A. Smith & A. Graham (eds.), *Destination London: The Expansion of the Visitor Economy* (pp. 117–139). London: University of Westminster Press.

15. Guerisoli F. (2018) Contemporary art and urban attractiveness: The role of hypermuseum and art foundations. In M. Paris (ed.), *Making Prestigious Places: How Luxury Influences the Transformation of Cities*. London: Routledge.
16. Guerisoli, 2018.
17. Guerisoli, 2018. Page 82.
18. MAXXI: National Museum of XXI Century Arts, Rome. Architect: Zaha Hadid, 2010.
19. Centre National d'Art et de Culture Georges Pompidou. Architects: Renzo Piano and Richard Rogers, 1977.
20. Acropolis Museum, Athens. Architects: Bernard Tschumi and Michael Photiadis, 2009.
21. Guerisoli, 2018.
22. Barreneche, R. A. (2005) *New Museums*. London: Phaidon.
23. Magnago Lampugnani, V., & Sachs, A. (eds.) (1999) *Museums for a New Millennium: Concepts Projects Buildings*. Munich: Prestel.
24. Barreneche, 2005. Page 6.
25. Van Moos, S. (1999) A museum explosion: Fragments of an overview. In V. Magnago Lampugnani & A. Sachs (eds.), *Museums for a New Millennium: Concepts Projects Buildings* (pp. 15–27). Munich: Prestel.
26. Dickenson Evitts, E. (2017) Louvre Pyramid: The folly that became a triumph. *Architect,* April 19. Available at: https://www.architectmagazine.com/awards/aia-honor-awards/louvre-pyramid-the-folly-that-became-a-triumph_o (accessed 8 February 2021).
27. Magnago Lampugnani & Sachs, 1999. Page 15.
28. Magnago Lampugnani & Sachs, 1999.
29. Lomholt, I. (2020) National Gallery London: Architecture. E-Architect. March 6. Retrieved: https://www.e-architect.com/london/national-gallery.
30. Jewish Museum, Berlin. Architect: Daniel Libeskind, 1999.
31. Phaeno Science Center, Wolfsburg. Architect: Zaha Hadid, 2005.
32. Graz Kunsthaus. Architects: Peter Cook and Colin Fournier, 2002
33. Imperial War Museum North, Manchester. Architect: Studio Daniel Libeskind, 2002.
34. Guggenheim Museum, Bilbao. Architect: Frank Gehry, 1997.
35. 'Case Study AT1: The redesign of the Military History Museum, Dresden' is sourced from: Ebejer, R. (2019) *An Exploration of Architectural Additions as a Means for Museum Redesign in Historic Buildings*. University of Malta (masters thesis).
36. Specht, 2014. Page 117.
37. Landry, C (2006) *The Art of City Making*. London: Earthscan. Page 148.
38. Sklair, L. (2006) Iconic architecture and capitalist globalization. *City* 10(1), 21–47.
39. Alaily-Mattar, N., Dreher, J., & Thierstein, A. (2018) Repositioning cities through star architecture: How does it work? *Journal of Urban Design* 23(2), 169–192.

40. Smith, A., & von Krogh Strand, I. (2011) Oslo's new Opera House: Cultural flagship, regeneration tool or destination icon? *European Urban and Regional Studies* 18(1), 93–110.
41. Alaily-Mattar et al., 2018.
42. Weidenfeld, A (2010) Iconicity and flagshipness of tourist attractions. *Annals of Tourism Research* 37(3), 851–854.
43. Jencks, C. (2006) The iconic building is here to stay. *City* 10(1), 3–20.
44. Hall & Barrett, 2018.
45. Marques & McIntosh, 2018.
46. Marques & McIntosh, 2018.
47. Jones, P. (2011) *The Sociology of Architecture: Constructing Identities.* Liverpool, UK: Liverpool University Press.
48. Sklair, 2006. Page 21.
49. Marques & McIntosh, 2018.
50. Hall & Barrett, 2018.
51. Smith & von Krogh Strand, 2011.
52. Konrad, D. (2010) Collecting the icon or semiotics of tourism. In J. Richter (ed.), *The Tourist City Berlin: Tourism and Architecture* (pp. 227–235). Salenstein: Braun.
53. Ponzini, D., Fotev, S., & Mavaacchio, F. (2016) Place making or place faking? The paradoxical effects of transnational circulation of architectural and urban development projects. In A. Russo and G. Richards (eds.), *Reinventing the Local in Tourism: Producing, Consuming and Negotiating Place* (pp. 153–170). Bristol: Channel View Publications.
54. Jencks, 2006.
55. Marques & McIntosh, 2018.
56. Marques & McIntosh, 2018.
57. Landry, 2006.
58. Smith & von Krogh Strand, 2011.
59. Guerisoli, 2018. Page 84.
60. Evans, G. (2003). Hard-branding the cultural-city: from Prado to Prada. *International Journal of Urban and Regional Research* 27(2), 417–440.
61. Evans, 2003. Page 431.
62. Forster, K. (1999) Frank O. Gehry: Guggenheim Museum, Bilbao, 1991–1997. In V. Magnago Lampugnani & A. Sachs (eds.), *Museums for a New Millennium: Concepts Projects Buildings* (pp. 124–131). Munich: Prestel.
63. Britannica (n.d.) Guggenheim Museum, Bilbao. Available at: https://www.britannica.com/topic/Guggenheim-Museum-Bilbao (accessed 2 February 2021).
64. Forster, 1999.
65. Ockman, J. (2004) New politics of the spectacle: 'Bilbao' and the global imagination. In D. Lasansky & B. McClaren (eds.), *Architecture and Tourism: Perception, Performance and Place* (pp. 227–240). Oxford, UK: Berg.
66. Specht, 2014.
67. Scerri et al., 2019.

68. Gravari-Barbas, M., Avila-Gómez, A., & Ruiz, D. C. (2018). Arquitectura, museos, turismo: La guerra de las marcas. *Revista de Arquitectura (Bogotá)* 20(1), 102–114.

69. Baniotopoulou, E. (2001) Art for whose sake? Modern Art Museums and their role in transforming societies: the case of the Guggenheim Bilbao. *Journal of Conservation and Museum Studies* 7, 1–5.

70. Plaza, B., & Haarich, S. (2015) The Guggenheim Museum Bilbao: between regional embeddedness and global networking. *European Planning Studies* 23(8), 1456–1475.

71. Hall & Barrett, 2018.

72. Ockman, 2004.

73. 'Case Study AT4: Norwegian National Opera and Ballet House, Oslo' is sourced from: Smith, A., & von Krogh Strand, I. (2011) Oslo's new Opera House: Cultural flagship, regeneration tool or destination icon? *European Urban and Regional Studies* 18(1), 93–110.

74. Smith & von Krogh Strand, 2011.

75. Smith & von Krogh Strand, 2011. Page 107.

76. 'Case Study AT5: The Temppeliaukio Church, Helsinki' is sourced from: Jokela, S. E. (2014) Tourism and identity politics in the Helsinki churchscape. *Tourism Geographies* 16(2), 252–269.

77. Temppeliaukio Church, Lutheran. Architects: Timo and Tuomo Suomalainen. 1969.

# CHAPTER 11

1. The industrial revolution was the transition to new manufacturing processes in Europe and the United States. The period spanned from approximately the mid-eighteenth century to the mid-nineteenth century. Instead of by hand, goods were being produced using machinery powered by steam or flowing water. There were also new chemical manufacturing and iron production processes. Factories were established in cities for the production of goods.

2. Shone, A., & Parry, B. (2004) *Successful Event Management: A Practical Handbook* (second edition). London: Continuum.

3. Bowdin, G., O'Toole, W., Allen, J., Harris, R., & McDonnell, I. (2006) *Events Management* (second edition). Oxford, UK: Elsevier.

4. Quinn, B. (2005) Arts festivals and the city. *Urban Studies* May, 42(5/6), 927–943.

5. Nolan, E. (2020) The supply and design of different types of venues for business events. *International Journal of Tourism Cities* 6(4), 691–710.

6. Johansson, M., & Kociatkiewicz, J. (2011) City festivals: Creativity and control in staged urban experiences. *European Urban and Regional Studies* 18(4), 392–405.

7. Smith, A. (2012) *Events and Urban Regeneration: The Strategic Use of Events to Revitalise Cities*. London: Routledge.

8. Smith, 2012.
9. Shone & Parry, 2004.
10. Frost, W. (2012) Events and tourism. In S. J. Page & J. Connell (eds.), *The Routledge Handbook of Events*. Abingdon: Routledge.
11. Getz, D. (2014). Timing tourism: MICE, events and mega events. In A. A. Lew, C. M. Hall, & A. M. Williams (eds.), *The Wiley Blackwell Companion to Tourism* (pp. 401–412). Oxford, UK: Wiley-Blackwell.
12. Getz, 2014.
13. Quinn, 2005.
14. Selby, M. (2004) *Understanding Urban Tourism: Image, Culture and Experience*. London: I. B. Tauris. Page 22.
15. Richards, G., & Palmer, R. (2010) *Eventful Cities: Cultural Management and Urban Revitalisation*. London: Routledge.
16. Johansson & Kociatkiewicz, 2011.
17. Ebejer, J. (2016) Regenerating Valletta: A vision for Valletta beyond 2020. In J. Ebejer (ed.), *Proceedings of Valletta Alive Foundation Seminar: Valletta Beyond 2020* (35–44). Available at: https://www.um.edu.mt/library/oar/handle/123456789/17857 (accessed 15 February 2021).
18. An 'auberge' was a sort of a monastery in which the Knights used to live. Knights of different nationalities used to have their own 'auberge' in Valletta and hence there were 'Auberge d' Italie', 'Auberge de Castille' and so on. The auberges that are still in existence are now used as government offices or museums.
19. Richards & Palmer, 2010.
20. Richards & Palmer, 2010.
21. Johansson & Kociatkiewicz, 2011.
22. Richards & Palmer, 2010. Page 72.
23. UNESCO (n.d.) *Intangible Cultural Heritage: Ommegang of Brussels, an Annual Historical Procession and Popular Festival*. Available from: https://ich.unesco.org/en/RL/ommegang-of-brussels-an-annual-historical-procession-and-popular-festival-01366 (accessed 5 February 2021).
24. UNESCO (n.d.) *Intangible Cultural Heritage: Flamenco*. Available from: https://ich.unesco.org/en/RL/flamenco-00363 (accessed 5 February 2021).
25. 'Case Study EV1: Il Palio di Siena' is sourced primarily from: Britannica (n.d.) The Palio–Italian festival. Available at https://www.britannica.com/sports/the-Palio (accessed 28 February 2021).
26. Getz, 2014.
27. Smith, 2012.
28. Getz, 2014.
29. Johansson & Kociatkiewicz, 2011.
30. Gregory, D. (2021, January 2) 350 Years of Rembrandt Showcased in Three Dutch Cities. *Museum Spotlight Europe*. Available at: https://museumspotlighteurope.com/350-years-of-rembrandt-showcased-in-three-dutch-cities/ (accessed 21 February 2021).
31. TEA/AECOM (2020) TEA/AECOM 2019 Theme Index and Museum Index: The Global Attractions Attendance Report. *Themed Entertainment Association*. Available at: https://aecom.com/theme-index/ (accessed 2 February 2021).

32. Frost, 2012.
33. Getz, D. (2007) *Event Studies: Theory, Research and Policy for Planned Events*. London: Butterworth-Heinneman.
34. Richards & Palmer, 2010.
35. Getz, D. (2013) *Event Tourism: Concepts, International Case Studies and Research*. Putnam Valley, NY: Cognizant Communications Corporation.
36. Place marketing and its role in urban development policies are discussed in chapter 2.
37. Johansson & Kociatkiewicz, 2011.
38. Quinn, 2005.
39. Frost, 2012.
40. Frost, 2012.
41. Getz, 2007.
42. Frost, 2012.
43. Richards & Palmer, 2010.
44. Richards & Palmer, 2010.
45. Thornley, A (2002) Urban Regeneration and Sports Stadia. *European Planning Studies* 10(7), 813–818.
46. Smith, 2012.
47. This aspect of city development is discussed in chapter 10.
48. Richards & Palmer, 2010.
49. Frost, 2012.
50. Newbold, C., & Jordan, J. (eds.) (2016) *Focus on World Festivals. Contemporary Case Studies and Perspectives*. Oxford, UK: Goodfellow Publishers.
51. Smith, 2012.
52. Smith, 2012.
53. Campelo, A. (2017) *Handbook on Place Branding and Marketing*. Cheltenham: Elgar.
54. Built in 1996 the Johan Cruyff Arena was originally known as Amsterdam ArenA. It was renamed in 2015 in honour of the legendary Dutch footballer.
55. Thornley, 2002.
56. 'Case study EV2: Galway Arts Festival' is sourced from: Quinn, B. (2005) Arts Festivals and the City. *Urban Studies*, May, 42(5/6), 927–943.
57. 'Case Study EV3: Stockholm Culture Festival' is sourced from: Johansson, M., & Kociatkiewicz, J. (2011) City festivals: Creativity and control in staged urban experiences. *European Urban and Regional Studies* 18(4), 392–405.
58. Immler, N., & Sakkers, H. (2014) (Re)Programming Europe: European capitals of culture: Rethinking the role of culture. *Journal of European Studies* 44(1), 3–29.
59. Immler & Sakkers, 2014.
60. Fox, T., & Rampton, J. (2019) *Ex-post Evaluation of the 2018 European Capitals of Culture: A Final Report by the EFECTIV Consortium*. Brussels: European Commission.
61. Richards, G., & Wilson, J. (2004) The impact of cultural events on city image: Rotterdam, Cultural Capital of Europe 2001. *Urban Studies* 41(10), 1931–1951.

62. Garcia, B. (2005) Deconstructing the city of culture: The long-term cultural legacies of Glasgow 1990. *Urban Studies* 42(5-6), 841–68.

63. Sassatelli, M. (2006) The logic of Europeanizing cultural policy. In U. Meinhof & A. Triandafyllidou (eds.), *Transcultural Europe: Cultural Policy in a Changing Europe* (pp. 24–42). Basingstoke: Palgrave Macmillan.

64. Griffiths R. (2006) City/culture discourses: Evidence from the competition to select the European Capital of Culture 2008. *European Planning Studies* 14(4), 415–430.

65. Smith, 2012.

66. Bianchini, F., Albano, R., & Bollo, A. (2013) The regenerative impacts of the European city/capital of culture events. In M. Leary & J. McCarthy (eds.), *The Routledge Companion to Urban Regeneration* (pp. 515–525). New York: Routledge.

67. Immler & Sakkers, 2014.

68. Ebejer, J., Xuereb, K. & Avellino, M. (2020) A critical debate of the cultural and social effects of Valletta 2018 European Capital of Culture. *Journal of Tourism and Cultural Change*, DOI: 10.1080/14766825.2020.1849240.

69. The bid book is the application which bidding cities submit and upon which the eventual European Capital of Culture is selected.

70. Žilič-Fišer, S., & Erjavec, K. (2015) The political impact of the European Capital of Culture: 'Maribor 2012 gave us the power to change the regime'. *International Journal of Cultural Policy* 23(5), 581–596.

71. Bianchini et al., 2013.

72. Sacco, P., & Tavano Blessi, G. (2007) European culture capitals and local development strategies: Comparing the Genoa 2004 and Lille 2004 cases. *Homo Oeconomicus* 24(1), 111–141.

73. Palmer, R. (2004). *European Cities and Capitals of Culture: Study Prepared for the European Commission, Part 1*. Brussels: Palmer-Rae Associates.

74. Richards & Wilson, 2004.

75. Garcia, 2005.

76. Johansson & Kociatkiewicz, 2011.

77. Spirou, C. (2011) *Urban Tourism and Urban Change: Cities in a Global Economy*. New York: Routledge.

78. Quinn, 2005. Page 932.

79. 'Case Study EV5: Lille 2004, European Capital of Culture' is sourced primarily from: Sacco & Tavano Blessi, 2007.

80. Richards, G., & Duif, L. (2019) *Small Cities with Big Dreams: Creative Placemaking and Brand Strategies*. New York: Routledge.

81. Richards & Duif, 2019.

82. Valletta is the capital of Malta. It is also a World Heritage Site. Malta joined the European Union in 2004.

83. 'Case Study EV6: Valletta 2018, European Capital of Culture' is sourced from: Ebejer et al. (2020).

84. Ebejer, J., & Tunbridge, J. (2020) The role of urban heritage in Malta's tourism: Issues and initiatives. In G. Cassar & M. Avellino (eds.), *Tourism and the Maltese Islands: Observations, Reflections and Proposals* (pp. 149–172).

Malta: University of Malta. Institute for Tourism, Travel and Culture and Kite Group.

85. The bid book is the application upon which Valletta was selected as European Capital of Culture. It is a commitment of what was to be done and achieved during the ECoC year.

86. In March 2018, the social media comments of V18 chairman appeared to mock the memory of Daphne Caruana Galizia, a Maltese investigative journalist murdered in October 2017. The comments were deemed to be highly offensive by a significant section of the population.

## CHAPTER 12

1. UNWTO (2018) *European Union Tourism Trends*. Available at: https://www.unwto.org/europe/publication/european-union-tourism-trends (accessed 15 January 2021).
2. Dodds, R. and Butler, R. (eds.). (2019) *Overtourism; Issues, Realities and Solutions*. Berlin: DeGruyter Studies in Tourism.
3. Milano, C., Cheer, J. M., & Novelli, M. (eds.). (2019) *Overtourism: Excesses, Discontents and Measures in Travel and Tourism*. Wallingford: CABI.
4. Séraphin, H., Gladkikh, T., & Thanh, T. V. (eds.). (2020) *Overtourism: Causes, Implications and Solutions*. London: Palgrave Macmillan.
5. González A., Fosse, J., and Santos-Lacueva, R. (2018) *The Integration of Sustainability in Tourism Policies of Major European Cities*. Barcelona City Council. Available at: http://www.ecounion.eu/wp-content/uploads/2018/11/The-integration-of-sustainability-in-tourism-policies-of-major-European-cities.pdf (accessed 30 January 2021).
6. Romagosa, F. (2020) The COVID-19 crisis: Opportunities for sustainable and proximity tourism. *Tourism Geographies* 2(3), 690–694.
7. Romagosa, 2020.
8. Speakman, M., & Sharpley, R., (2012) A chaos theory perspective on destination crisis management: evidence from Mexico. *Journal of Destination Marketing and Management* 1, 67–77.
9. Le, D., & Phi, G. (2020) Strategic responses of the hotel sector to COVID-19: Toward a refined pandemic crisis management framework. *International Journal of Hospitality Management*. DOI: https://doi.org/10.1016/j.ijhm.2020.102808.
10. Improved connectivity of cities is discussed in chapter 3.
11. Solaris, J. (2021, January 20) Companies focus on how to maximize revenue from virtual events. *EventMB*. Available at: https://www.eventmanagerblog.com/companies-focus-on-virtual-events (accessed 17 February 2021).
12. Buckley, J. (2021, March 2) Italy has a new way to combat overtourism. CNN Travel. Available at: https://edition.cnn.com/travel/article/uffizi-diffusi-tuscany-galleries-overtourism/index.html (accessed 15 March 2021).
13. The 'albergo diffuso' is discussed in chapter 6.
14. TEA/AECOM, 2020.

15. European Commission. (n.d.) *Energy, Climate change, Environment*. Available at: https://ec.europa.eu/info/energy-climate-change-environment_en (accessed 28 February 2021).

16. Higgins-Desbiolles, F. (2020) The end of global travel as we know it: an opportunity for sustainable tourism. *The Conversation*. Available at: https://theconversation.com/the-end-of-global-travel-as-we-know-it-an-opportunity-for-sustainable-tourism-133783 (accessed 28 February 2021).

17. Romagosa, 2020.

18. Knežević Cvelbar, L., & Ogorevc M. (2020) Saving the tourism industry with staycation vouchers. *Emerald Open Research* 2(65).

19. The Future of Tourism Coalition. (n.d.) *Guiding Principle 8: Mitigate Climate Impacts*. Available at: https://www.futureoftourism.org/principle-8 (accessed 12 March 2021).

## CHAPTER 13

1. Ashworth, G., & Tunbridge, J. (2000) *The Tourist-Historic Cities: Retrospect and Prospect of Managing the Heritage City*. London: Routledge.

2. Wall, G., & Mathieson, A. (2005) *Tourism: Change, Impacts and Opportunities*. Harlow: Pearson.

3. Ritchie, J., & Crouch, G. (2003). *The Competitive Destination: A Sustainable Tourism Perspective*. Cambridge, MA: CABI.

# Bibliography

Alaily-Mattar, N., Dreher, J., & Thierstein, A. (2018) Repositioning cities through star architecture: How does it work? *Journal of Urban Design* 23(2), 169–192.

Andersson Cederholm, E. (2004) The use of photo-elicitation in tourism research-framing the backpacker experience. *Scandinavian Journal of Hospitality and Tourism* 4(3), 225–241.

Anhar, L. (2001) *The definition of boutique hotels*. Available at: www.hospitalitynet.org/news/4010409.html (accessed 18 June 2018).

Archdaily (2017, February 1) Diana, Princess of Wales Memorial Fountain. *ArchDaily*. Available at: https://www.archdaily.com/803509/diana-princess-of-wales-memorial-fountain-gustafson-porter-plus-bowman (accessed 8 November 2020).

Ashworth, G. (1995) Managing the cultural tourist. In G. Ashworth & A. Doetvorst (eds.), *Tourism and Spatial Transformations*. Wallingford: CABI.

Ashworth, G., & Bruce, D. (2009) Town walls, walled towns and tourism: Paradoxes and paradigms. *Journal of Heritage Tourism* 4(4), 299–313.

Ashworth, G., & Page, S. (2011) Urban tourism research: Recent progress and current paradoxes. *Tourism Management* 32, 1–15.

Ashworth G. J., & Tunbridge, J. E. (2000) *The Tourist-Historic Cities: Retrospect and Prospect of Managing the Heritage City*. London: Routledge.

Ashworth, G., & Tunbridge, J. (2004) *Malta Makeover: Prospects for the Realignment of Heritage, Tourism and Development*. URSI Report 304, Urban and Regional Studies Institute, University of Groningen.

Assaeroporti (n.d.) Dati Annuali. Available at: https://assaeroporti.com/dati-annuali/ (accessed 10 December 2020).

Bandarin, F., & Van Oers, R. (2012) *The Historic Urban Landscape: Managing Heritage in an Urban Century*. Chichester: Wiley-Blackwell.

Baniotopoulou, E. (2001) Art for whose sake? Modern Art Museums and their role in transforming societies: The case of the Guggenheim Bilbao. *Journal of Conservation and Museum Studies* 7, 1–5.

Barreneche, R. A. (2005) *New Museums*. London: Phaidon.

BBC (2011) *Museums Enjoy 10 Years of Freedom*. Available at: https://www.bbc.com/news/entertainment-arts-15927593 (accessed 3 January 2021).

Benfield, R. (2018) Garden tourism. In S. Agarwal, G. Busby & R. Huang (eds.), *Special Interest Tourism: Concepts, Contexts and Cases* (pp. 156–170). Wallingford: CABI.

Bergenhenegouwen, G., & van Weesep, J. (2003) Manipulated space: The Beurstraverse retail complex in Rotterdam. *Belgeo* 1, 79–86.

Berleant, A. (2003) The aesthetic in place. In S. Menin (ed.), *Constructing Place: Mind and Matter* (pp. 41–54). London: Routledge.

Bianchini, F., Albano, R., & Bollo, A. (2013) The regenerative impacts of the European City/capital of culture events. In M. Leary & J. McCarthy (eds.), *The Routledge Companion to Urban Regeneration*. New York: Routledge.

Boisen, M., Terlouw, K., Groote, P., & Couwenberg, O. (2018) Reframing place promotion, place marketing, and place branding: Moving beyond conceptual confusion. *Cities* 80, 4–11.

Booth, P. (2010) Sheffield: A miserable disappointment no more? In J. Punter (ed.), *Urban design and the British urban renaissance* (pp. 85–99). London: Routledge.

Bowdin, G., O'Toole, W., Allen, J., Harris, R., & McDonnell, I. (2006) *Events Management* (second edition). Oxford, UK: Elsevier.

Britannica. (n.d.) *Botanical Garden*. Available at: https://www.britannica.com/science/botanical-garden-study-and-exhibition-garden (accessed 15 January 2021).

Britannica. (n.d.) *Guggenheim Museum, Bilbao*. Available at: https://www.britannica.com/topic/Guggenheim-Museum-Bilbao (accessed 2 February 2021).

Britannica (n.d.) *The Palio-Italian Festival*. Available at https://www.britannica.com/sports/the-Palio (accessed 28 February 2021).

Bruce, D. & Creighton, O. (2006) Contested identities: The dissonant heritage of European town walls and walled towns. *International Journal of Heritage Studies* 12(3), May, 234–254. Page 240.

Buckley, J. (2021, March 2) Italy has a new way to combat overtourism. *CNN Travel*. Available at https://edition.cnn.com/travel/article/uffizi-diffusi-tuscany-galleries-overtourism/index.html (accessed 15 March 2021).

Burgen, S. (2020, October 20) La Rambla: Plans to transform Barcelona's tourist rat run into a cultural hub. *Guardian*. Available at: https://www.theguardian.com/travel/2020/oct/20/la-rambla-transform-barcelona-tourist-rat-run-into-a-cultural-hub (accessed 12 December 2020).

Burns, A. (2000) Emotion and urban experience: Implications for design. *Design Issues* 16(3), 67–79.

Cadell, C., Falk, F., & King, F. (2008) *Regeneration in European Cities: Making Connections*. York: Joseph Rowntree Foundation.

Calle-Vaquero, M.d.l., García-Hernández, M., & Mendoza de Miguel, S. (2021) Urban planning regulations for tourism in the context of overtourism: Applications in historic centres. *Sustainability* 13, 70.

Campelo, A. (2017) *Handbook on Place Branding and Marketing*. Cheltenham: Elgar.

Campelo, A. (2017) The state of the art: From country-of origin to strategies for economic development. In A. Campelo (ed.), *Handbook on Place Branding and Marketing* (pp. 3–21). Cheltenham: Elgaronline.

Campisi, D., Costa, R., & Mancuso, P. (2010) The effects of low cost airlines growth in Italy. *Modern Economy* 1, 59–67.
Canepa, S., & Ab Ghafar, N. (2020) Water in architecture, architecture of water. *Journal of Civil Engineering and Architecture* 14, 249–262.
Canter, D. (1977) *The Psychology of Place*. London: Architectural Press.
Carmona, M., Heath, T., Oc, T., & Tiesdall, S. (2010) *Public Places - Urban Spaces: The Dimensions of Urban Design* (second edition). Oxford, UK: Architectural Press.
Castello, L. (2010) *Rethinking the Meaning of Place: Conceiving Place in Architecture-Urbanism*. Farnham: Ashgate.
Cellini, R., & Cuccia, T. (2018) How free admittance affects charged visits to museums: An analysis of the Italian case. *Oxford Economic Papers* 70(3), 680–698.
Chronis, A. (2012) Tourists as story-builders: Narrative construction at a heritage museum. *Journal of Travel and Tourism Marketing* 29(5), 444–459.
Ciesluk, K. (2020, June 14) *How Do Low-Cost Carriers Actually Make Money: A Complete Breakdown*. Available at https://simpleflying.com/how-low-cost-carriers-make-money/ (accessed 10 December 2020).
Cohen, E. (1979) A phenomenology of tourist experiences. *Sociology* 13, 179–201.
Colomb, C. (2012) *Staging the New Berlin: Place Marketing and the Politics of Urban Reinvention Post-1989*. London: Routledge.
Connell, J., & Meyer, D. (2004) Modelling the visitor experience in the gardens of Great Britain. *Current Issues in Tourism* 7(3), 183–216.
Correira Loureiro, S., Rita, P., & Moraes Sarmento, E. (2020) What is the core essence of small city boutique hotels? *International Journal of Culture, Tourism and Hospitality Research* 14(1), 44–62.
Couch, C., Sykes, O., & Cocks, M. (2013) The changing context of urban regeneration in North West Europe. In M. E. Leary & J. McCarthy (eds.), *The Routledge Companion to Urban Regeneration* (pp. 33–44). New York: Routledge.
Crang, M. (1997) Picturing practices: Research through the tourist gaze. *Progress in Human Geography* 21(3), 359–373.
Crespi-Vallbona, M., & Pérez, M. D. (2015) Tourism and food markets: A typology of food markets from case studies of Barcelona and Madrid. *Regions Magazine* 299(1), 15–17.
Creswell, J. W. (2007) *Qualitative Inquiry and Research Design: Choosing among Five Traditions* (second edition). Thousand Oaks, CA: Sage.
Curtis, S. (2019) The river Thames: London's Riparian highway. In A. Smith and A. Graham (eds.), *Destination London: The Expansion of the Visitor Economy* (pp. 165–182). London: University of Westminster Press.
Cuthbert, A. (2006) *The Form of Cities*. Oxford, UK: Blackwell.
D'Aveni, R. (2010) How to escape the differentiation proliferation trap. *Strategy and Leadership* 38(3), 44–49.
Davidson, R., & Cope, B. (2003) *Business Travel: Conference, Incentive Travel, Exhibitions, Corporate Hospitality and Corporate Travel*. Harlow: Pearson Education Limited.
Dick, M., Eveleigh, D., & Sullivan, J. (2019) *The Black Country: A History in 100 Objects*. West Midlands: Black Country Living Museum Publications.

Dickenson Evitts, E. (2017, April 19) Louvre Pyramid: The folly that became a triumph. *Architect*. Available at: https://www.architectmagazine.com/awards/aia-honor-awards/louvre-pyramid-the-folly-that-became-a-triumph_o (accessed 8 February 2021).

Dimelli, D. (2019) Modern conservation principles and their application in Mediterranean Historic Centres: The case of Valletta. *Heritage* 2, 787–796.

Dimitrovskia D., & Crespi-Vallbona, M. (2018) Urban food markets in the context of a tourist attraction: La Boqueria market in Barcelona. *Tourism Geographies* 20(3), 397–417.

Dodds, R., & Butler R. (eds.). (2019) *Overtourism: Issues, Realities and Solutions*. Berlin: DeGruyter Studies in Tourism.

Dunne, G., Flanagan, S., & Buckley, J. (2010) Towards an understanding of International City Break Travel. *International Journal Tourism Research* 12, 409–417.

Ebejer, J. (2015) *Tourist Experiences of Urban Historic Areas: Valletta as a Case Study* (doctoral thesis). University of Westminster.

Ebejer, J. (2016) Regenerating Valletta: A vision for Valletta beyond 2020. In J. Ebejer (ed.), *Proceedings of Valletta Alive Foundation Seminar: Valletta beyond 2020* (pp. 35–44). Available at: https://www.um.edu.mt/library/oar/handle/123456789/17857 (accessed 15 February 2021).

Ebejer, J. (2019) Using fortifications for tourism: Can conservation objectives be reconciled with financial sustainability? In R. Kusek & J. Purchla (eds.), *Heritage and Society* (pp. 353–366). Krakow: International Cultural Centre.

Ebejer, J., & Dimelli, D. (2019) *Conservation Issues of Two Fortified Historic Towns and World Heritage Sites: Rhodes and Valletta*. Conference paper at 5th Heritage Forum of Central Europe, International Cultural Centre, Krakow Poland. 19–20 September.

Ebejer, J., Smith, A., Stevenson, N., & Maitland R. (2020) The tourist experience of heritage urban spaces: Valletta as a case study. *Tourism Planning and Development* 17(4), 458–474.

Ebejer, J., & Tunbridge, J. (2020) The role of urban heritage in Malta's tourism: Issues and initiatives. In G. Cassar & M. Avellino (eds.), *Tourism and the Maltese Islands: Observations, Reflections and Proposals* (pp. 149–172). Malta: University of Malta. Institute for Tourism, Travel and Culture and Kite Group.

Ebejer, J., Xuereb, K., & Avellino, M. (2020) A critical debate of the cultural and social effects of Valletta 2018 European Capital of Culture. *Journal of Tourism and Cultural Change*, DOI: 10.1080/14766825.2020.1849240.

Ebejer, R (2019) *An Exploration of Architectural Additions as a Means for Museum Redesign in Historic Buildings* (masters thesis). University of Malta.

Eco, U. (1986) Function and sign: Semiotics of architecture. In M. Gottdiener & A. Lagopoulos (eds.), *The City and the Sign*. Guildford, NY: Columbia University Press.

Edensor, T. (2000) Staging tourism: Tourists as performers. *Annals of Tourism Research* 27(2), 322–344.

Edwards, D., Griffin, T., & Hayllar, B. (2008) Urban tourism research: Developing an agenda. *Annals of Tourism Research* 35(4), 1032–1052.

Ek, R., Larsen, J., Hornskov, S., & Mansfeldt, O. (2008) A dynamic framework of tourist experiences: Space-time and performances in the experience economy. *Scandinavian Journal of Hospitality and Tourism* 8(2), 122–140.

European Commission. (n.d.) *Energy, Climate Change, Environment.* Available at: https://ec.europa.eu/info/energy-climate-change-environment_en (accessed 28 February 2021).

European Commission (n.d.) Transport and mobility: Trans-European Transport Network (TEN-T). Available at: https://ec.europa.eu/transport/themes/infrastructure/ten-t_en (accessed 28 February 2021).

Evans, G. (2003) Hard-branding the cultural-city: From Prado to Prada. *International Journal of Urban and Regional Research* 27(2), 417–440.

Fletcher, J., Fyall, A., Gilbert, D., & Wanhill S (eds.). (2013) *Tourism: Principles and Practice* (fifth edition). London: Pearson.

Forster, K. (1999) Frank O. Gehry: Guggenheim Museum, Bilbao, 1991–1997. In V. Magnago Lampugnani & A. Sachs (eds.), *Museums for a New Millennium: Concepts Projects Buildings* (pp. 124–131). Munich: Prestel.

Fox, T., & Rampton, J. (2019) *Ex-post Evaluation of the 2018 European Capitals of Culture. A Final Report by the EFECTIV Consortium.* Brussels: European Commission.

Frankl, V. (1992) *Man's Search for Meaning: An Introduction to Logotherapy.* London: Rider.

Frost, W. (2012) Events and tourism. In S. J. Page & J. Connell (eds.), *The Routledge Handbook of Events.* Abington: Routledge.

Füller, H., & Michel B. (2014) "Stop Being a Tourist!" new dynamics of urban tourism in BerlinKreuzberg. *International Journal of Urban and Regional Research* 38, 1304–1318.

Garcia, B. (2005) Deconstructing the city of culture: The long-term cultural legacies of Glasgow 1990. *Urban Studies* 42(5–6), 841–868.

Garrod B (2009) Understanding the relationship between tourism destination imagery and tourist photography. *Journal of Travel Research* 47(3), 346–358.

Getz, D. (2007) *Event Studies: Theory, Research and Policy for Planned Events.* London: Butterworth-Heinneman.

Getz, D. (2013) *Event Tourism: Concepts, International Case Studies and Research.* Putnam Valley, NY: Cognizant Communications Corporation.

Getz, D. (2014) Timing tourism: MICE, events and mega-events. In A. A. Lew, C. M. Hall, & A. M. Williams (eds.), *The Wiley Blackwell Companion to Tourism* (pp. 401–412). Oxford, UK: Wiley-Blackwell.

Getz, D. (2016) *Event Studies: Theory, Research and Policy for Planned Events* (third edition). London: Routledge.

Gilli, M., & Ferrari, S. (2016) The 'Diffuse Hotel': An Italian new model of sustainable hospitality. In A. Russo & G. Richards (eds.), *Reinventing the Local in Tourism: Producing, Consuming and Negotiating Place* (pp. 65–83). Bristol: Channel View Publications.

Gilli, M., & Ferrari, S. (2018) Tourism in multi-ethnic districts: The case of Porta Palazzo market in Torino. *Leisure Studies* 37(2), 146–157.

Goeldner, C., & Ritchie, J. R. B. (2009). *Tourism: Principles, Practices, Philosophies.* Hoboken, NJ: Wiley.

González A., Fosse, J., & Santos-Lacueva, R. (2018) *The Integration of Sustainability in Tourism Policies of Major European Cities.* Barcelona City Council. Available at: http://www.ecounion.eu/wp-content/uploads/2018/11/The-integration-of-sustainability-in-tourism-policies-of-major-European-cities.pdf (accessed 30 January 2021).

Gordon, I., & Buck, N. (2005) Introduction: Cities in the new conventional wisdom. In N. Buck, I. Gordon, A. Harding, & I. Turok (eds.), *Changing Cities; Rethinking Urban Competiveness, Cohesion and Governance* (pp. 1–21). London: Palgrave Macmillan.

Gospodini, A. (2009) Post-industrial trajectories of mediterranean European cities: The case of post-Olympics Athens. *Urban Studies* 46(5-6), 1157–1186.

Gospodini, A. (2001) Urban design, urban space morphology, urban tourism: An emerging new paradigm concerning their relationship. *European Planning Studies* 9(7), 925–934.

Governing Body of Suomenlinna. (2000) *Suomenlinna: Conservation and Reuse.* Helsinki, Finland.

Governing Body of Suomenlinna. (2015) *A sustainable tourism strategy for Suomenlinna.* Helsinki, Finland.

Governing Body of Suomenlinna. (2012) *At fort: Self-analysis report.* Helsinki, Finland.

Graham, B., Ashworth, G., & Tunbridge, J. (2016) *A Geography of Heritage.* London: Routledge.

Grauman, C. F. (2002) *The phenomenological approach to people-environment studies.* In R. Bechtel, & A. Churchman (eds.), *Handbook of Environmental Psychology* (pp.95–113). New York: Wiley.

Gravari-Barbas, M., Avila-Gómez, A., & Ruiz, D. C. (2018). Arquitectura, museos, turismo: La guerra de las marcas. *Revista de Arquitectura (Bogotá)* 20(1), 102–114.

Gregory, D. (2021, January 2) *350 Years of Rembrandt Showcased in Three Dutch Cities Museum Spotlight Europe.* Available at: https://museumspotlighteurope.com/350-years-of-rembrandt-showcased-in-three-dutch-cities/ (accessed 21 February 2021).

Griffin, T., & Hayllar, B. (2007) Historic waterfronts as tourism precincts: An experiential perspective. *Tourism and Hospitality Research* 7(1), 3–16.

Griffin, T., & Hayllar, B. (2009) Urban tourism precincts and the experience of place. *Journal of Hospitality Marketing and Management* 18(2-3), 127–153.

Griffiths, A. (2014, March 22) Rotterdam Centraal station reopens with a pointed metal-clad entrance. *DeZeen.* Available at: https://www.dezeen.com/2014/03/22/rotterdam-centraal-station-benthem-crouwel-mvsa-architects-west-8/ (accessed 18 February 2021).

Griffiths, R. (2006) City/culture discourses: Evidence from the competition to select the European Capital of Culture 2008. *European Planning Studies* 14(4), 415–430.

*Guardian* (2000, 12 June) *Swaying Millennium Bridge to close after two days.* Available at: https://www.theguardian.com/uk/2000/jun/12/2 (accessed 12 February 2021).
Guerisoli, F. (2018) Contemporary art and urban attractiveness: The role of hyper-museum and art foundations. In M. Paris (ed.), *Making Prestigious Places; How Luxury Influences the Transformation of Cities* (81–92). London: Routledge.
Gustafson, P. (2001) Meanings of place: Everyday experience and theoretical conceptualizations. *Journal of Environmental Psychology* 21(1), 5–16.
Guttentag, D. (2015) Airbnb: Disruptive innovation and the rise of an informal tourism accommodation sector. *Current Issues in Tourism* 18(12), 1192–1217.
Haldrup, M., & Larsen, J. (2003) The family gaze. *Tourist Studies* 3(1), 23–45.
Hall, C. M., & Page, S. J. (2014) *The Geography of Tourism and Recreation. Environment, Place and Space* (fourth edition). London: Routledge.
Hall, C. M., & Ram, Y. (2019) Measuring the relationship between tourism and walkability? Walk Score and English tourist attractions. *Journal of Sustainable Tourism* 27(2), 223–240.
Hall, T. (1997) *Planning Europe's Capital Cities: Aspects of Nineteenth Century Urban Development.* London: E&FN Spon.
Hall, P. (1998) *Cities in Civilization.* London: Weidenfeld and Nicholson.
Hall, T., & Barrett, H. (2018) *Urban Geography* (fifth edition). London: Routledge.
Hannabus, S. (1999) Postmodernism and the heritage experience. *Library Management* 20(5), 295–302.
Hass-Klau, C. (1990) *The Pedestrian and City Traffic.* London: Belhaven Press.
Hayllar, B., & Griffin, T. (2007) A tale of two precincts. In J. Tribe & D. Airey (eds.), *Developments in Tourism Research* (155–169). Oxford, UK: Elsevier.
Henderson, J. (2018) Making cities more walkable for tourists: A view from Singapore's streets. *International Journal of Tourism Cities* 4(3), 285–297.
Higgins-Desbiolles, F. (2020) The end of global travel as we know it: An opportunity for sustainable tourism. *The Conversation.* Available at: https://theconversation.com/the-end-of-global-travel-as-we-know-it-an-opportunity-for-sustainable-tourism-133783 (accessed 28 February 2021).
Hiss, T. (1991) *The Experience of Place.* New York: Knopf.
Holcomb, B. (1999) Marketing cities for tourism. In D. R. Judd & S. S. Fainstein (eds.), *The Tourist City* (54–70). New Haven, CT: Yale University Press.
Holloway, C. J., & Taylor, N. (2006) *The Business of Tourism* (seventh edition). Harlow: Financial Times & Prentice Hall.
Hughes, K., Bond, N., & Ballantyne, R. (2013) Designing and managing interpretive experiences at religious sites: Visitors' perceptions of Canterbury Cathedral. *Tourism Management* 36, 210–220.
Immler, N., & Sakkers, H. (2014) (Re)Programming Europe: European capitals of culture: Rethinking the role of culture. *Journal of European Studies* 44(1), 3–29.
Jackson, S. (2018) London Mithraeum: Reimagining the famous Roman temple. *Current Archaeology* 334, December 1.
Jamal, T., & Hollinshead, K. (2001) Tourism and the forbidden zone: The underserved power of qualitative inquiry. *Tourism Management* 22, 63–82.

Jencks, C. (2006) The iconic building is here to stay. *City* 10(1), 3–20.

Jiven, J., & Larkham, P. (2003) Sense of place: Authenticity and character. *Journal of Urban Design* 8(1), 67–82.

Johansson, M., & Kociatkiewicz, J. (2011) City festivals: Creativity and control in staged urban experiences. *European Urban and Regional Studies* 18(4), 392–405.

Jokela, S. E. (2014) Tourism and identity politics in the Helsinki churchscape. *Tourism Geographies* 16(2), 252–269.

Jones, D., Day, J., & Quadri-Felitti, D. (2013) Emerging definitions of boutique and lifestyle hotels: A Delphi study. *Journal of Travel & Tourism Marketing* 30(7), 715–731.

Jones, P. (2011) *The Sociology of Architecture: Constructing Identities*. Liverpool, UK: Liverpool University Press.

Judd, D. (1999) Constructing the tourist bubble. In D. Judd & S. Fainstein (eds.), *The Tourist City* (pp. 35–53). New Haven, CT: Yale University Press.

Keyser, H. (2002) *Tourism Development*. New York: Oxford University Press.

Kim, H. J. (2019) *Suomenlinna, enjoy with care: Designing for visitor guidance service with behavioural insights* (master thesis). Aalto University.

Knežević Cvelbar, L., & Ogorevc, M. (2020) Saving the tourism industry with staycation vouchers. *Emerald Open Research* 2(65).

Kolb, B. (2017) *Tourism Marketing for Cities and Towns: Using Social Media and Branding to Attract Tourists*. London: Routledge.

Konrad, D. (2010) Collecting the icon or semiotics of tourism'. In J. Richter (ed.), *The Tourist City Berlin: Tourism and Architecture* (pp. 227–235). Salenstein: Braun.

Kovářová, L. (2017) *UNESCO World Heritage Sites: Ways of Presenting and Interpreting the Past* (PhD thesis). Charles University in Prague.

Kyle, G. T., Mowen A. J., & Tarrant, M. (2004) Linking place preferences with place meaning: An examination of the relationship between place motivation and place attachment. *Journal of Environmental Psychology* 24(4), 439–454.

Landry, C. (2006) *The Art of City Making*. London: Earthscan.

Lassen, C., Smink, C., & Smidt-Jensen, S. (2009) Experience spaces, (aero)mobilities and environmental impacts. *European Planning Studies* 17(6), 887–903.

Law, C. (2002) *Urban Tourism: The Visitor Economy and the Growth of Large Cities* (second edition). New York: Continuum.

Law, C. M. (2000) Regenerating the city centre through leisure and tourism. *Built Environment* 26, 117–129.

Le, D., Phi, G. (2020) Strategic responses of the hotel sector to COVID-19: Toward a refined pandemic crisis management framework. *International Journal of Hospitality Management*. DOI: https://doi.org/10.1016/j.ijhm.2020.102808.

Leask, A. (2008) The nature and role of visitor attractions. In A. Fyall, B. Garrod, A. Leask & S. Wanhill (eds.), *Managing Visitor Attractions: New Directions* (pp. 3–15) (second edition). Oxford, UK: Butterworth-Heinemann.

Lee, D. H. (2010) Digital cameras, personal photography and the reconfiguration of spatial experiences. *Information Society* 26(4), 266–275.

Leiper, N. (2004) *Tourism Management* (third edition). Sydney: Pearson Education Australia.

Leiper, N., & Park, S. Y. (2010) Skyscrapers' influence on cities' roles as tourist destinations. *Current Issues in Tourism* 13(4), 333–349.

Lengkeek, J. (2001) Leisure experience and imagination: Rethinking Cohen's modes of tourism experience. *International Sociology* 16(2), 173–184.

Lewis, G. D. (2020, November 6) Museum. *Encyclopædia Britannica*. Available at: https://www.britannica.com/topic/museum-cultural-institution (accessed 1 January 2021).

Lim, W. M., & Endean, M. (2009) Elucidating the aesthetic and operational characteristics of UK boutique hotels. *International Journal of Contemporary Hospitality Management* 21(1), 38–51.

Loeffler, T. (2004) A Photo elicitation study of the meanings of outdoor adventure experiences. *Journal of Leisure Research* 36(4), 536–556.

Lomholt, I. (2020, March 6) National Gallery London: Architecture. *E-Architect*. Available at: https://www.e-architect.com/london/national-gallery (accessed 15 January 2021).

Lorentzen, A. (2009) Cities in the experience economy. *European Planning Studies* 17(6), 829–845.

Lupu, C., Brochado, A., & Stoleriu, O. (2019) Visitor experiences at UNESCO monasteries in Northeast Romania. *Journal of Heritage Tourism* 14(2), 150–165.

Lynch, K. (1960) *The Image of the City*. Cambridge, MA: MIT Press.

MacLaran, A. (ed.). (2003) *Making Space: Property Development and Urban Planning*. London: Routledge.

Macleod, S., Hourston Hanks, L., & Hale, J. (2012) *Museum Making: Narrative, Architectures, Exhibitions*. London: Routledge.

Magnago Lampugnani, V., & Sachs, A. (eds.) (1999) *Museums for a New Millennium: Concepts Projects Buildings*. Munich: Prestel.

Maitland, R. (2008) Conviviality and everyday life: The appeal of new areas of London for visitors. *International Journal of Tourism Research* 10, 15–25.

Maitland, R. (2007) Culture, city users and the creation of new tourism areas in cities. In M. K. Smith (ed.), *Tourism, Culture and Regeneration*. Wallingford: CABI.

Maitland, R. (2006) How can we manage the tourist-historic city? Tourism strategy in Cambridge, UK, 1978–2003. *Tourism Management* 27(6), 1262–1273.

Maitland, R., & Newman, P. (2014) Developing world tourism cities. In R. Maitland & P. Newman (eds.), *World Tourism Cities: Developing Tourism Off the Beaten Track* (pp. 1–21). London: Routledge.

Manley, S., & Guise, R. (1998) Conservation in the built environment. In C. Greed & M. Roberts (eds.), *Introducing Urban Design: Interventions and Responses* (pp. 64–86). London: Routledge.

Marques, B., & McIntosh, J. (2018) The spell of the visual and the experience of the sensory: Understanding icons in the built environment. *Charrette* 5(1) Spring.

Mason, P. (2000) Zoo tourism: The need for more research. *Journal of Sustainable Tourism* 8(4), 333–339.

Mazumdar, Sh., and Mazumdar, Sa. (2004) Religion and place attachment: A study of sacred places. *Journal of Environmental Psychology* 24(3), September, 385–397.

McIntosh, A. J. (1999). Into the tourist's mind: Understanding the value of the heritage experience. *Journal of Travel and Tourism Marketing* 8(1), 41–64.

McKercher, B., & du Cros. (2002) *Cultural Tourism: Partnership between Tourism and Cultural Heritage Management.* Binghamton, NY: Haworth.

Milano, C., Cheer, J. M., & Novelli, M. (eds.). (2019) *Overtourism: Excesses, Discontents and Measures in Travel and Tourism.* Wallingford: CABI.

Millar, P. (1995) Heritage management for heritage tourism. In S. Medlik (ed.), *Managing Tourism* (second edition, pp. 115–238). Oxford, UK: Butterworth-Heinemann.

Montanari, F., Scapolan, A., & Codeluppi, E. (2013) Identity and social media in an art festival. In A. Munar, S. Gyimóthy, & L. Cai (eds.), *Tourism Social Media: Transformations in Identity, Community and Culture* (pp. 207–225). Bingley, UK: Emerald Group Publishing.

Montgomery, J. R. (1998) Making a city: Urbanity, vitality and urban design. *Journal of Urban Design* 3(1), 93–116.

Müller, S., & Brunner-Sperdin, A. (2006) Entertainment, attraction and engagement: Pre-condition for the new holiday experience and leisure time! In K. Weiermair, P. Pechlaner, & T. Bieger (eds.), *Time Shift, Leisure and Tourism: Impacts of Time Allocation on Successful Products and Services* (pp. 249–265). Berlin: Erich Schmidt Verlag GmbH & Co.

Newbold, C., & Jordan, J. (eds.) (2016) *Focus on World Festivals. Contemporary Case Studies and Perspectives.* Oxford, UK: Goodfellow Publishers.

Nientied, P. (2020) Rotterdam and the question of new urban tourism. *International Journal of Tourism Cities.* DOI: 10.1108/IJTC-03-2020-0033.

Nolan, E. (2020) The supply and design of different types of venues for business events. *International Journal of Tourism Cities* 6(4), 691–710.

Norberg-Schulz, C. (1988) *Architecture - Meaning and Place: Selected Essays.* New York: Rizzoli.

Norberg-Schulz, C. (1980) *Genius Loci: Towards a Phenomenology of Architecture.* New York: Rizzoli.

O'Mara, S. (2019) *In Praise of Walking: The New Science of How We Walk and Why It's Good for Us.* London: The Bodley Head.

Ockman, J. (2004) New politics of the spectacle: 'Bilbao' and the global imagination. In D. Lasansky & B. McClaren (eds.), *Architecture and Tourism: Perception, Performance and Place* (pp. 227–240). Oxford, UK: Berg.

Orbasli, A., (2000) *Tourists in Historic Towns: Urban Conservation and Heritage Management.* London: E&FN Spon.

Oriade, A., & Cameron, H. (2018) Cathedral tourism. In S. Agarwal, G. Busby, & R. Huang (eds.), *Special Interest Tourism: Concepts, Contexts and Cases* (pp. 40–51). Wallingford: CABI.

Öystilä, M., & Grönroos, L. (2017) Tourism in the world heritage site. In K. Havas (ed.), *Changes in the Hospitality Industry.* Helsinki: Haaga-Helia University of Applied Sciences.

Page, S. (2011) *Tourism Management: An Introduction* (fourth edition). London: Elsevier.

Palmer, R. (2004). *European Cities and Capitals of Culture: Study Prepared for the European Commission, Part 1.* Brussels: Palmer-Rae Associates.

Pasquinelli, C. (2017) Tourism connectivity and spatial complexity: A widening bi-dimensional arena of urban tourism research. In N. Bellini, & C. Pasquinelli (eds.), *Tourism in the City. Towards an Integrative Agenda on Urban Tourism* (pp. 29–50). Switzerland: Springer.

Pavia, A. (2018) Walking as a luxury activity. In M. Paris (ed.), *Making Prestigious Places: How Luxury Influences the Transformation of Cities*. London: Routledge.

Pendlebury, J. (2002) Conservation and regeneration: Complementary or conflicting processes? The case of Grainger Town, Newcastle upon Tyne. *Planning Practice and Research* 17(2), 145–158.

Penner, R., Adams, L., & Robinson, S. (2013) *Hotel Design Planning and Development* (second edition). New York: W. W. Norton & Company.

Pérez Guilarte, Y., & Lois González, R. C. (2018) Sustainability and visitor management in tourist historic cities: The case of Santiago de Compostela, Spain. *Journal of Heritage Tourism* 13(6), 489–505.

Pine, B., & Gilmore, J. (1999) *The Experience Economy*. Boston: Harvard Business School Press.

Plaza, B., & Haarich, S. (2015) The Guggenheim Museum Bilbao: Between regional embeddedness and global networking. *European Planning Studies* 23(8), 1456–1475.

Ponzini, D., Fotev, S., & Mavaacchio, F. (2016) Place making or place faking? The paradoxical effects of transnational circulation of architectural and urban development projects. In A. Russo & G. Richards (eds.), *Reinventing the Local in Tourism: Producing, Consuming and Negotiating Place* (pp. 153–170). Bristol: Channel View Publications.

Prideaux, B. (2009) *Resort Destinations: Evolution, Management and Development*. London: Routledge.

Prideaux, B., & Coghlan, A. (2010) Reef tour operators: Digital cameras and photo taking behaviour on the Great Barrier Reef: Marketing opportunities for Reef tour operators. *Journal of Vacation Marketing* 16, 171–183.

Punter, D. (2019, 12 February) Ulysses. *Encyclopedia Britannica*. Available at: https://www.britannica.com/topic/Ulysses-novel-by-Joyce (accessed 21 March 2021).

Punter, J. (1991) Participation in the design of urban space. *Landscape Design* 200, 24–27.

Quinn, B. (2005) Arts festivals and the city. *Urban Studies*, 42(5/6), 927–943.

Rabbiosi, C. (2015) Renewing a historical legacy: Tourism, leisure shopping and urban branding in Paris. *Cities* 42, 195–203.

Ragheb, M. (1996) The search for meaning in leisure pursuits: Review conceptualisation and a need for a psychometric development. *Leisure Studies* 15, 245–258.

Relph, E. (1976) *Place and Placelessness*. London: Pion Limited.

Richter, J. (ed.) (2010) *The Tourist City Berlin: Tourism and Architecture*. Salenstein: Braun.

Richards, G. (2014) Cultural tourism 3.0: The future of urban tourism in Europe? In R. Garibaldi (ed.), *Il turismo culturale europeo. Città ri-visitate. Nuove idee e forme del turismo culturale* (pp. 25–38). Milan: Franco Angeli.

Richards, G. (2016) Hostels and the making of urban spaces. In P. Russo and G. Richards (eds.), *Reinventing the Local in Tourism: Producing, Consuming and Negotiating Place* (pp. 171–184). Bristol: Channel View Publications.

Richards, G. (2017) Tourists in their own city: Considering the growth of a phenomenon. *Tourism Today* 16, 8–16.

Richards, G., & Duif, L. (2019) *Small Cities with Big Dreams: Creative Placemaking and Brand Strategies*. New York: Routledge.

Richards, G., & Marques, L. (2018) *Creating Synergies between Cultural Policy and Tourism for Permanent and Temporary Citizens*. Available at: http://www.agenda-21culture.net (accessed 15 February 2021).

Richards, G., & Palmer, R. (2010) *Eventful Cities: Cultural Management and Urban Revitalisation*. London: Routledge.

Richards, G., & Wilson, J. (2004) The impact of cultural events on city image: Rotterdam, cultural capital of Europe 2001. *Urban Studies* 41(10), 1931–1951.

Rickly Boyd, J. (2009) The tourist narrative. *Tourist Studies* 9(3), 259–280.

Rickly-Boyd, J., & Metro-Roland, M. (2010) Background to the fore: The prosaic in tourist places. *Annals of Tourism Research* 37(4), 1164–1180.

Ritchie, J., & Crouch, G. (2003). *The Competitive Destination: A Sustainable Tourism Perspective*. Cambridge, MA: CABI.

Roberts, P. (2000) The evolution, definition and purpose of urban regeneration. In P. Roberts & H. Sykes (eds.), *Urban Regeneration* (pp. 9–36). London: Sage.

Rodriguez, L. (2011, January 3) Pedestrian-only shopping streets make communities more livable. *Planetizen*. Available at: https://www.planetizen.com/node/47517 (accessed 20 February 2021).

Rogers, T. (1998) *Conferences: A Twenty-First Century Industry*. Harlow: Addison Wesley Longman.

Romagosa, F. (2020) The COVID-19 crisis: Opportunities for sustainable and proximity tourism. *Tourism Geographies* 2(3), 690–694.

Russell, D. (2003) Selling bradford: Tourism and northern image in the late twentieth century. *Contemporary British History* 17(2), 49–68.

Rutes, W., Penner, R., & Adams, L. (2001) *Hotel Design: Planning and Development*. New York: W. W. Norton & Company.

Sacco, P. L. (2011) *Culture 3.0: A New Perspective for the EU 2014–2020 Structural Funds Programming*. Report for the OMC Working Group on Cultural and Creative Industries.

Sacco, P., & Tavano Blessi, G. (2007) European culture capitals and local development strategies: Comparing the Genoa 2004 and Lille 2004 cases. *Homo Oeconomicus* 24(1), 111–141.

Samarasekara, G. N., Fukahori, K., & Kubota, Y. (2011) Environmental correlates that provide walkability cues for tourists: An analysis based on walking decision narrations. *Environment and Behaviour* 43(4), 501–524.

Sancar, F., & Severcan, Y. (2010) Children's places: Rural-urban comparisons using participatory photography in the Bodrum Peninsula, Turkey. *Journal of Urban Design* 15(3), 293–324.

Santos, J. R. (2019). Public space, tourism and mobility: Projects, impacts and tensions in Lisbon's urban regeneration dynamics. *Journal of Public Space* 4(2), 29–56. DOI: 10.32891/jps.v4i2.1203.

Sassatelli, M. (2006) The logic of Europeanizing cultural policy. In U. Meinhof & A. Triandafyllidou (eds.), *Transcultural Europe: Cultural Policy in a Changing Europe*, 24–42. Switzerland: Springer.

Scerri, M., Edwards, D., & Foley, C. (2019). Design, architecture and the value to tourism. *Tourism Economics* 25(5), 695–710.

Selby, M. (2004) *Understanding Urban Tourism: Image, Culture and Experience*. London: I. B. Tauris.

Séraphin, H., Gladkikh, T., & Thanh, T. V. (eds.) (2020) *Overtourism: Causes, Implications and Solutions*. London: Palgrave Macmillan.

Shackley, M. (1996) *Wildlife Tourism*. London: Routledge.

Shaw, G., & Williams, A. (2002) *Critical Issues in Tourism: A Geographical Perspective* (second edition). Oxford, UK: Blackwell.

Shone, A., & Parry, B. (2004) *Successful Event Management: A Practical Handbook* (second edition). New York: Continuum.

Shoval, N. (2018) Urban planning and tourism in European cities. *Tourism Geographies* 20(3), 371–376.

Shydlovska, K. (2019) *Outdooractive: Multimedia Fountain*. Available at: https://www.outdooractive.com/en/poi/wroclaw/multimedia-fountain/41228551/ (accessed 8 November 2020).

Sinha, A., & Sharma, Y. (2009) Urban design as a frame for site readings of heritage landscapes: A case study of Champaner-Pavagadh, Gujarat, India. *Journal of Urban Design* 14(2), 203–221.

Sklair, L. (2006) Iconic architecture and capitalist globalization. *City* 10(1), 21–47.

Smaldone, D., Harris, C., & Sanyal, N. (2005) An exploration of place as a process: The case of Jackson Hole, WY. *Journal of Environmental Psychology* 25(4), December, 397–414.

Smith, A. (2019) Conceptualising the expansion of destination London: Some conclusions. In A. Smith & A. Graham (eds.), *Destination London: The Expansion of the Visitor Economy* (pp. 225–236). London: University of Westminster Press.

Smith, A. (2019) Event takeover? The commercialisation of London's Parks. In A. Smith & A. Graham (eds.), *Destination London: The Expansion of the Visitor Economy* (pp. 205–224). London: University of Westminster Press.

Smith, A. (2012) *Events and Urban Regeneration: The Strategic Use of Events to Revitalise Cities*. London: Routledge.

Smith, A. (2019) Vertical city tourism: Heightened aesthetic and kinaesthetic experiences. In A. Smith & A. Graham (eds.), *Destination London: The Expansion of the Visitor Economy* (pp. 117–139). London: University of Westminster Press.

Smith, A., & von Krogh Strand, I. (2011) Oslo's new Opera House: Cultural flagship, regeneration tool or destination icon? *European Urban and Regional Studies* 18(1), 93–110.

Smørvik, K. (2020) Why enter the church on holiday? Tourist encounters with the Basilica of Santa Maria in Trastevere, Rome. *Journal of Heritage Tourism*. DOI: 10.1080/1743873X.2020.1807557.

Solaris, J. (2021, January 20) Companies focus on how to maximize revenue from virtual events. *EventMB*. Available at: https://www.eventmanagerblog.com/companies-focus-on-virtual-events (accessed 17 February 2021).

Southworth, M. (2005) Designing the walkable city. *Journal of Urban Planning and Development* 4(1), 246–257.

Speakman, M., & Sharpley, R., (2012) A chaos theory perspective on destination crisis management: Evidence from Mexico. *Journal of Destination Marketing and Management* 1, 67–77.

Specht, J. (2014) *Architectural Tourism: Building for Urban Travel Destinations*. Wiesbaden: Springer Gabler.

Spirou, C. (2011) *Urban Tourism and Urban Change: Cities in a Global Economy*. Routledge.

Staiff, R. (2012) The somantic and the aesthetic: Embodied heritage tourism experinces of Luang Prabang, Lao. In L. Smith, E. Waterton, & S. Watson (eds), *The Cultural Moment in Tourism* (pp. 38–55). London: Routledge.

Staniewska, A. (2019) Translating fortified landscape heritage to the public: Dilemmas on rehabilitation, popularization and conservation methods. In R. Kusek & J. Purchla (eds.), *Heritage and Society* (pp. 189–204). Krakow: International Cultural Centre.

Stevens, Q. (2009) Nothing more than feelings. *Architectural Theory Review* 14(2), 156–172.

Stevenson, N., & Farrell, H. (2017) Taking a hike: Exploring leisure walkers embodied experiences. *Social and Cultural Geography* 19(4), 429–447.

Stokowski, P. (2002) Languages of place and discourses of power: Constructing new senses of place. *Journal of Leisure Research* 34(4), 368–382.

Strike, J. (1994) *Architecture in Conservation: Managing Developments at Historic Sites*. Oxon: Routledge.

Suvantola, J. (2002) *Tourist's Experience of Place*. Aldershot: Ashgate.

Tallon, A. (2013) *Urban Regeneration in the UK* (second edition). London: Routledge.

TEA/AECOM (2020) TEA/AECOM 2019 *Theme Index and Museum Index: The Global Attractions Attendance Report*. Themed Entertainment Association. Available at: https://aecom.com/theme-index/ (accessed 12 December 2020).

The Future of Tourism Coalition (n.d.) *Guiding principle 8: Mitigate climate impacts*. Available at https://www.futureoftourism.org/principle-8 (accessed 12 March 2021).

Thornley, A. (2002) Urban regeneration and sports stadia. *European Planning Studies* 10(7), 813–818.

Timothy, D. (2005) *Shopping, Tourism, Retail and Leisure*. Clevedon: Channel View Publications.

Timothy, D. (1997) Tourism and the personal heritage experience. *Annals of Tourism Research* 24(3), 751–753.

Timothy, D. (2014) Trends in tourism, shopping and retailing. In A. Lew, M. Hall & A. Williams (eds.), *The Wiley Blackwell Companion to Tourism* (pp. 378–388). Oxford, UK: Wiley Blackwell.

Timothy, D., & Boyd, S. (2015) *Tourism and Trails: Cultural, Ecological and Management Issues*. Bristol: Channel View Publications.

Trew, J., & Cockerell, N. (2002) The European market for UK city breaks. *Insights* 14(58), 85–111.

Trueman, M., Cook, D., & Cornelius, N. (2008) Creative dimensions for branding and regeneration: Overcoming negative perceptions of a city. *Place Branding and Public Diplomacy* 4(1), 29–44.

Tuan, Y. F. (1977) *Space and Place: The Perspective of Experience*. Minneapolis: University of Minnesota Press.

Tunbridge, J. (2018) Thirty years on have we resolved the heritage questions? *Heritage Quarterly* 32–33, 131–141.

Tunbridge, J., & Ashworth, G. (1996) *Dissonant Heritage: Managing the Past as a Resource in Conflict*. Chichester: John Wiley.

Tweed, C., & Sutherland, M. (2007) Built cultural heritage and sustainable urban development. *Landscape and Urban Planning* 83(1), 62–69.

Uggla, Y. (2014) Protecting urban greenery: The case of Stockholm's National City Park. *City & Community*, December, 13(4).

Uggla, Y., & Olausson, U. (2013) The enrollment of nature in tourist information: Framing urban nature as 'the other'. *Environmental Communication* 7(1), 97–112.

UNESCO. (n.d.) *Intangible Cultural Heritage: Flamenco*. Available from https://ich.unesco.org/en/RL/flamenco-00363 (accessed 5 February 2021).

UNESCO. (n.d.) *Intangible Cultural Heritage: Ommegang of Brussels, an Annual Historical Procession and Popular Festival*. Available from: https://ich.unesco.org/en/RL/ommegang-of-brussels-an-annual-historical-procession-and-popular-festival-01366 (accessed 5 February 2021).

UNESCO World Heritage Convention. (n.d.) *World Heritage List*. Available from: https://whc.unesco.org/en/list/ (accessed 15 November 2020).

UNWTO. (2018) *European Union Tourism Trends*. Available at: https://www.unwto.org/europe/publication/european-union-tourism-trends (accessed 15 January 2021).

UNWTO World Tourism Organization and ETC European Travel Commission. (2011) *Handbook on Tourism Product Development*. Madrid: WTO.

Urry, J. (2002) *The Tourist Gaze* (second edition). London: Sage.

Vahtikari, T. (2003) Urban interpretations of world heritage: Re-defining the city. In Marjaana Niemi, & Ville Vuolanto (eds.), *Reclaiming the City: Innovation, Culture, Experience* (pp. 123–164). Helsinki: Studia Fennica Historica.

Vahtikari, T. (2017) *Valuing World Heritage Cities*. London: Routledge.

Van Dijck, J. (2008) Digital photography: Communication, identity, memory. *Visual Communication* 7(1), 57–76.

Van Moos, S. (1999) A museum explosion: Fragments of an overview. In V. Magnago Lampugnani & A. Sachs (eds.), *Museums for a New Millennium: Concepts Projects Buildings* (pp. 15–27). Munich: Prestel.

VisitBritain. (n.d.) *Official Statistics*. Available at: https://www.visitbritain.org/official-statistics (accessed 10 December 2020).

Voase, R. (2008) Rediscovering the imagination: Meeting the needs of the 'new' visitor. In A. Fyall, B. Garrod, A. Leask, & S. Wanhill (eds.), *Managing Visitor Attractions: New Directions* (second edition). Oxford, UK: Butterworth-Heinemann.

Wall, G., & Mathieson, A. (2005) *Tourism: Change, Impacts and Opportunities*. Harlow: Pearson.

Wang, N. (1999) Rethinking authenticity in tourism experience. *Annals of Tourism Research* 26(2), 349–370.

Wearing, B., & Wearing, S. L. (1996) Refocusing the tourist experience: the flaneur and the chorister. *Leisure Studies* 1, 229–243.

Wearing, S. L., & Wearing, B. (2001) Conceptualizing the selves of tourism. *Leisure Studies* 20(2), 143–159.

Weddle Landscape Design. (n.d.) *Sheffield Winter Gardens*. Available at: http://weddles.co.uk/portfolio/sheffield-winter-gardens/ (accessed 16 January 2021).

Weidenfeld, A. (2010) Iconicity and flagshipness of tourist attractions. *Annals of Tourism Research* 37(3), 851–854.

Westwood, S. (2006) Shopping in sanitised and un-sanitised spaces. Adding value to the tourist experiences. *Journal of Retail and Leisure Property* 5(4), 281–291.

Wiki2. (n.d.) *List of the Busiest Airports in the Nordic Countries*. Available at: https://wiki2.org/en/List_of_the_largest_airports_in_the_Nordic_countries#2019_statistics (accessed 15 January 2021).

Winter, M., & Gasson, R. (1996) Pilgrimage and tourism: Cathedral visiting in contemporary England. *International Journal of Heritage Studies* 2(3), 172–182.

Winter, T. (2013) Cultures of interpretation. In R. Staiff, R. Bushell, & S. Watson, (eds.), *Heritage and Tourism: Place, Encounter, Engagement* (pp. 172–186). London: Routledge.

Woodward, R. (2005) Water in landscape. In H. Dreiseitl & D. Grau (eds.), *New Waterscapes: Planning, Building and Designing with Water* (pp. 23–29). Basel: Birkauser.

Woodward, S. (2004) Faith and tourism: Planning tourism in relation to places of worship. *Tourism and Hospitality Planning and Development* 1(2), 173–186.

World Cities Ranking. (n.d.) *Largest and Best Aquariums in Europe 2021*. Available at: https://worldcitiesranking.com/largest-and-best-aquariums-in-europe-top-10/ (accessed 26 February 2021).

Wunderlich, F. M. (2008) Walking and rhythmicity: Sensing urban space. *Journal of Urban Design* 13(1), 125–139.

Žilič-Fišer, S., & Erjavec, K. (2015) The political impact of the European Capital of Culture: 'Maribor 2012 gave us the power to change the regime'. *International Journal of Cultural Policy* 23(5), 581–596.

# Index

Aarhus, 45; ARoS Art Museum, 175
agriculture, 11
air travel, 9, 41, 201, 206
Airbnb. *See* short-term rental tourism accommodation
airport hotels, 97
airports, 41
albergo diffuso (diffuse hotel), 94, 204
Amsterdam, 15, 29, 132, 167, 188, 194, 199; Johan Cruyff Arena, 192; Rijksmuseum, 188
amusement arcades, 89
animal captivity, 87
Antwerp, Havenhuis, 176
archeological sites, 74
architectural icons, 21, 173, 174, 209
architecture, 6, 15, 17, 18, 20, 97, 134–36, 143, 161, 165–79, 207
art galleries, 22, 29, 48, 66, 104
Ashworth, Gregory, 2
Athens: 2004 Olympic Games, 191; Acropolis, 62, 110, 111; Acropolis Museum, 168; Plaka, 62
authenticity, 137, 138, 140, 156
aviation, 41, 205
Avignon, Festival d'Avignon, 182

Banská Štiavnica, 160
Barcelona, 12, 15, 199; Barri Gòtic, 40, 125; Casa Batlló, 166; Ciutat Vella, 125; El Raval, 125; Hospital de la Santa Creu i Sant Pau, 125; La Barceloneta, 125; La Boqueria market, 78; La Princesa market, 78; La Ribera, 125; La Sagrada Familia, 40; La Sagrada Família, 125; Las Ramblas, 31, 56, 78; MACBA in Raval, 168; Museu Nacional d'Art de Catalunya, 125; Olympic Games, 1992, 20, 126; Parc Güell, 40; Plaça d' Europa, 125; Plaça de Carles Buïgas, 125; Plaça de l'Odissea, 125; Plaça de les Cascades, 125; Plaça de l'Ictineo, 125; Port Olímpic, 40; Rambla de Mar, 125; Sagrada Familia, 80; Santa Caterina market, 78; Teatre Principal, 31
Bath, 102
Bayreuth, Bayreuth Festival, 181
Belfast, Titanic Museum, 176
Bergamo, 42; Orio Al Serio Airport, 42
Berlin, 29, 165, 167, 170, 171, 176; Berlin wall, 15; Potsdamer Platz, 56; Tropical Islands, 89
bidding for major events, 182
Bilbao, Guggenheim Museum, 166, 167, 171, 177
Billund, 44; Billund Airport, 45, 46; Lalandia Billund, 45; Legoland Billund, 44, 46

Birmingham: Black Country Living Museum, 75; Cadbury World, 76
Blackpool, Blackpool Pleasure Beach, 88, 89
Bologna, 43
botanical gardens, 84
Boulogne-sur-Mer, Nausicaá, 88
boutique hotels, 95, 96
Bradford, 19, 147; Bradford Arts Festival, 147
Bratislava, 43
Brescia, 42
Brussels, 44; Ommegang, 186
built environment, 6, 207
business events, 183
business travel, 2, 93, 201

car parks, 121
Cardiff, 189
castles, 67
catering establishments, 165, 207
catering facilities, 18, 48, 99
cathedrals, 66
Chester, 102
churches, 66, 77, 155
city attractiveness, 16, 29, 31, 60, 83, 150, 165, 168
city authorities, 14, 19, 23, 26, 49, 202, 208
city breaks, 38
city centres, 12, 13, 93, 94, 98, 99, 119, 122
climate change, 205
Cologne, Cologne Cathedral, 81
Commonwealth Games, 183
competitive cities, 16, 20, 27, 49, 207
concert halls, 13
concerts, 34, 48, 52
conference centres, 16, 20, 103
conferences, 2, 34, 48, 202
contemporary architecture, 165–67, 171, 175, 176
Copenhagen, 88; Bakken, 88; Langelinie, 128; Little Mermaid, 128, 129; Tivoli Gardens, 88, 89

Cork, 195
COVID-19 pandemic, 199, 209
COVID-safety, 40, 203, 204
Cricket World Cup, 183
crisis situations in tourism, 200
cultural activities, 196, 203
cultural events, 183, 187, 188, 195
cultural facilities, 30, 38
culture, 30
cycling, 120

De Efteling, Netherlands, 89
degrowth strategies, 205
Denmark, 46
deprived urban areas, 30
design hotels, 97
destination management, 56, 208
destination management organisations, 50
development control, 25
docks, 11
Dresden, 148, 161; Military History Museum, 171, 176
Dublin, 15, 195; General Post Office, 167; Guinness Storehouse, 76, 77; Temple Bar, 29; Ulysses walk, 79
Dubrovnik, 73

Edinburgh, 161; Edinburgh Castle, 110; Edinburgh International Festival, 182
Elvas, 73
entertainment, 15
environmental stimuli, 131
Esbjerg, 45
escape rooms, 89
Essen-Ruhr, 195
ethnic cuisines, 99
ethnic heritage, 17
EU funding, 42, 102, 155–58
Europa-Park, Germany, 89
European Capital of Culture, 183, 193–98
European culture, 31
European Economic Community, 194
European Regional Development Funds, 155–57

European Travel Commission, 51
event facilities, 190, 191, 196
events, 20, 30, 49, 52, 60, 85, 103, 135, 163, 168, 181–98, 202, 209
exhibition venues, 102
exhibitions, 48, 104, 202

factories, 11
Fado, 31
festivalisation, 182, 184, 190
festivals, 20, 181–90, 192
FIFA World Cup, 182
financial services, 29
flight ticket prices, 42, 199
Florence, 204; Ponte Vecchio, 102; Uffizi Gallery, 204
food and drink establishments. *See* restaurants
food markets, 77, 78
Formula 1 Grand Prix, 183
fortifications, 73, 155, 162
fortified towns, 66
forward planning, 25
Fredericia, 45
Future of Tourism Coalition, 205

Galway, Galway Arts Festival, 192
Gaudí, Antoni, 40
Gdańsk, 43
Gdynia, 43
Gehry, Frank, 177
*genius loci*. *See* sense of place
Genoa, 195
Germany, 19
Glasgow, European Capital of Culture, 1990, 196
global economic crisis of 2008/2009, 199
global warming, 204
globalisation, 14
glocalisation, 190
Gothenburg, Norra Älvstranden, 29
governance, 4, 5, 14, 31
Granada, 73
Graz, 43, 161; Kunsthaus, 171, 176

Greece, 191
Guggenheim effect, 178
Guimarães, 195

The Hague, 132, 188
hallmark event, 183
Hamburg, Deichtorhallen, 170
health issues, 200, 204
Hellevoetsluis, 73
Helsinki: Ehrensvard Museum, 163; Governing Body of Suomenlinna, 162–64; Olympic Games, 1952, 163; Suomenlinna, 162–64; Suomenlinna masterplan, 164; Temppeliaukio Church, 175, 179
herd immunity, 201
heritage experience, 136
historic areas, 15, 21, 23, 31, 66, 74, 77, 80, 96, 108, 123, 136, 139, 140, 146, 149, 150, 209
historic buildings, 66, 74, 81, 82, 96, 151, 156, 196
historic towns, 133, 145–47, 152, 154, 155, 160
Horsens, 45
hotel chains, 93
hotels, 13, 91, 93, 94, 165, 203, 207
household goods, 11
Hull, The Deep, 176
hybrid events, 202
hypermuseum, 168

identity, 18, 19, 21, 38
impacts of tourism, 3
India, 181
industrial revolution, 12, 75, 181
industrialisation, 12, 26, 159, 181
Innsbruck, Swarovski Kristallwelten, 76
Intangible Cultural Heritage, 185, 186
inter-city coach services, 43
International Council on Monuments and Sites, 160
investor confidence, 30
Ironbridge Gorge, 75

Jerusalem, 165
Jutland, Denmark, 45

Katowice, 43
Kolding, 45
Koper, 43
Košice, 195
Kraków, 15; Wawel Cathedral, 80; Wawel Royal Castle, 48

land transport, 42, 44
landmarks, 133
Lego, 44
Leiden, 188
leisure, 137, 208
leisure settings, 56, 57
Libeskind, Daniel, 171
lifestyles, 37, 40
light rail, 121
Lille, 197; European Capital of Culture, 2004, 197; Maisons Folie, 197; Roubaix, 29
Lisbon: European City of Culture 1994, 32; MAAT-Museum of Art, Architecture and Technology, 32; Museum of Aljube, 32; Oceanário de Lisboa, 88; Tagus River, 31; UEFA European Championship 2004, 32; urban regeneration, 31; World Exposition 1998, 32
Liverpool, 62, 161
Ljubljana, 43
London, 2, 44; ArcelorMittal Orbit, 168; Bankside, 29, 116, 117, 143; Blakes Hotel, 95; Covent Garden, 102; Crystal Palace, 12, 181; Festival of Britain, 1951, 182; Great Exhibition, 1851, 12, 181; Islington, 143; Jubilee Park, 117; London Aquarium, 117; London Docklands, 27; London Eye, 116, 117, 167; London Mithraeum, 75; Madame Tussuads, 137; Millennium Bridge, 116; National Theatre, 117; Notting Hill Carnival, 185, 190; O2 Arena, 168; Olympic Park, 168; Royal Botanic Gardens, Kew, 84; Royal Festival Hall, 116, 117, 182; Sainsbury Wing of the National Gallery, 170; Sherlock Holmes museum, 79; SkyGarden, 167; South Bank, 29, 116, 117, 182; St. Paul's Cathedral, 116, 117; Tate Modern, 117, 167; View from the Shard, 167
low-cost airlines, 37, 41
Luxembourg, 195
Lynch, Kevin, 132

Madrid: San Fernando market, 78; San Miguel market, 78; Santiago Bernabéu Stadium, 79
Malta, 42, 183, 185, 197, 198; Cittadella, Gozo, 73, 157; Cospicua, 73, 156, 158; Cottonera, 73, 79, 157; EU membership, 157; Fort St. Angelo, 157; fortifications, 156, 158; Grand Harbour, 66, 73, 156, 157; Great Siege, 1565, 140; Ħaġar Qim and Mnajdra temples, 74; Marsamxett Harbour, 158; Mdina, 73, 79; Rabat, Gozo, 157; role in World War II, 141, 157; Senglea, 73, 156; Valletta. *See* Valletta; Vittoriosa, 73, 156
Manchester: Imperial War Museum North, 171; Old Trafford, 79
Mantova, 42
manufacturing, 11–14
Maribor, 43
marina hotels, 97
market research, 18
Marseille, 44
meaning, 133–40
Mediterranean coast, 41, 203, 206
Mediterranean culture, 31
mental maps, 132
Mercouri, Melina, 193
MICE. *See* business events
Milano, 42, 132; Linate Airport, 41; Malpensa Airport, 41; Stadio Giuseppe Meazza, 79

military architecture, 156, 158, 162
model railways, 89
monasteries, 67, 79, 104
monuments, 149
movement through space, 132
Munich, 15; city centre, 120; Nymphenburg Palace, 84; Olympic Games, 1972, 120; Therme Erding, 89
museums, 16, 20, 29, 38, 40, 48, 168–70, 175, 177, 203, 208
mystery, 111, 143

Naples, Pompeii, 110
narrative, 15, 66, 81, 139, 140
National Eisteddfod, Wales, 189
National Energy and Climate Plans, 205
national parks, 67
natural coastlines, 67
natural disasters, 200
natural environment, 6
nature reserves, 67, 88
Netherlands, 19, 43, 188

Olympic Games, 182
Oslo: Bjørvika, 178, 179; Norwegian National Opera House, 178
overtourism, 199, 205

package tours, 41
Padova, 43
Pakistan, 181
Paris, 12, 44, 167; Beaubourg, 176, 177; Centre Pompidou, 170, 175, 177; Centres Georges Pompidou, 168; Disneyland Park, 89; Eiffel Tower, 12, 128, 136, 172; Exposition Universelle, 1889, 12; FIFA World Cup, 1998, 190; La Défense, 56; Musée d'Orsay, 170; Notre Dame Cathedral, 79; St. Denis, 190; Stade de France, 190
Paris Agreement, 205
park and ride service, 121
*passegiata*, 107

Patras, 195
Pécs, 195
pedestrian streets, 15, 119–23
pedestrianisation. *See* pedestrian streets
pedestrians, 108
photography, 127–30
physical environment, 2
Piano, Renzo, 176
Piran, 73
Pisa, Tower of Pisa, 174
place branding, 16
place marketing, 16, 18, 19
place selling, 17
Poznań, 43
Prague: Aquapalace, 89; Charles Bridge, 102; Old Market Square, 101; Royal Mile, 101
pride, sense of, 19
Prince Charles, 170
public gardens, 66
public transport, 30, 94, 120, 121, 125, 183

railways, 12, 43, 93
Rauma, 160
Ravenna, 43
religious pilgrimage, 138
Rembrandt, 188
residential use, 13, 23, 24, 39, 56, 63, 73, 162, 163, 165
restaurants, 13, 99, 100, 203
retailing, 12, 13
Rhodes, 73
Riga, 161
road infrastructure, 43
Rogers, Richard, 176
Rome, 11, 132, 165; Basilica of Santa Maria, 80; churches in Rome, 80; Fontana di Trevi, 56, 58; MAXXI National Museum of XXI Century Arts, 168; St. Peter's Basilica, 79, 166
Rotterdam, 132; Erasmus Bridge, 44; Koopgoot, 122; Kop van Zuid, 29; Nieuwe Maas, 44;

Rotterdam Centraal station, 43–45; Stationsplein, 44
Rugby World Cup, 183
rural villages, 67

Salzburg: Dreifaltigkeitskirche, 83; Franziskanerkirche, 83; Kajetanerkirche, 83; Kollegienkirche, 83; Markuskirche, 83; Michaelskirche, 83; Mozart's house, 79; Old Town (Altstadt), 82; Salzburg Cathedral, 82; Salzburger Festspiele, 181; St. Peter's Abbey, 82; St. Sebastian Church and Cemetery, 83
San Gimigniano, 73, 166
sandy beaches, 67
seasonality, 49
Second World War, 147
self-catering apartments, 93
sense of place, 131, 133–35, 139, 143
Sheffield: Castle Square, 33; Don Valley Athletics Stadium, 34; Meadowhall shopping centre, 34; Millennium Galleries, 34; Peace Gardens, 33; Ponds Forge Sports Complex, 34; Sheffield Hallam University, 34; Sheffield Winter Gardens, 84–86; Supertram, 34; University of Sheffield, 34
shopping, 2, 48, 101, 102, 122, 208
short-term rental tourism accommodation, 40, 97
Sibiu, 195
Siena: Corteo Storico, 186; Il Palio, 185, 186
sightseeing, 10, 67, 108, 110, 208
skyscrapers, 167
Slovenia, 205
small businesses, 50, 202, 208
Spain, 199
sports complexes, 104
sports events, 183, 191, 208
sports stadia, 16, 22, 183
star architects, 174
stately homes, 67, 71

Stavanger, 195
Stockholm: Djurgarden, 85; Eco Park Association, 85; Lidingö, 85; National City Park, 85; Solna, 85; Stockholm Culture Festival, 193; Stockholm University, 87; Stockholm Water Festival, 193
Strasbourg, 15
street furniture, 123
streetscapes, 109, 143
sustainability, 44
Świnoujście, 43
Szczecin, 43

technology, 11, 12
terrorist activity, 200
theatres, 38, 52
theme parks, 52, 66, 87–89
timber balconies, 155
Torino, Porta Palazzo market, 78
tourism accommodation, 48, 91
tourism demand, 49
tourism experience, 135, 137, 138
tourism facilities, 49, 168
tourism planning, 3
tourism product, 18, 47, 49, 50
tourism product development, 19, 51, 208
tourism resources, 4, 47
tourist accommodation, 18
tourist experience, 3, 50, 134, 138, 139
tourist motivation, 8, 111, 208
tourist-historic cities, 149
trade, 11
tram systems. *See* light rail
Trans-European Transport Network (TEN-T), 42
transport infrastructure, 6, 48, 94
travel insurance, 201
Trento, Castello del Buonconsiglio, 72
Trieste, 42, 43
Tunbridge, John, 2
Turku, 195
Tuscany, 204

Udine, 43
Umeå, 195
UNESCO World Heritage Committee, 159, 160
university campuses, 103
urban aesthetics, 22
urban areas, 26
urban conservation, 148, 149, 151–57
Urban Conservation Areas, 153
urban design, 6
urban heritage, 78, 108, 110, 112, 118, 145–48, 150, 153, 154, 156
urban parks, 6, 17, 18
urban planning, 5, 12, 24, 25, 27, 135, 153
urban regeneration, 26, 28, 29, 150, 190, 195
urban spaces, 17, 19–21, 110, 126, 183, 184, 196, 207, 209
urban tourism, 1–3, 5, 39, 108, 208
urbanisation, 12, 14, 159, 181
Urbino, 160

vaccination, 201
Valencia, Oceanogràfic, 88
Valletta, 77, 140, 155, 158, 160, 185, 197, 198; City Gate, 176; European Capital of Culture, 2018, 198; Fort St. Elmo, 157; Fortifications Interpretation Centre, 158; Merchants Street, 124; Notte Bianca, 185; Parliament Building, 176; St. James Cavalier, 157; St. John's Co-Cathedral, 48, 80; Upper Barrakka Gardens, 66
Vejle, 45
Venice, 43, 160; Murano, 76; Ponte di Rialto, 56
Verona, 42
Vienna, 12, 15, 29, 43, 161; Prater, 89; Schönbrunn Palace, 84
viewing platforms in cities, 168
Vilnius, 161
vineyards, 67
virtual events, 202
vision, 131
visiting friends and relatives, 2, 208
visitor attractions, 8, 13, 15, 18, 65–67, 70, 79, 82, 84, 88, 89, 104, 165, 203, 207
visitor experience. *See* tourist experience

Wales, 76, 189
walkability, 113–16, 118
walking, 18, 94, 107, 109, 117, 209
walking trails, 18, 67, 118
walled towns, 73, 155, 158
Warsaw, 43, 148
water, 57
water parks, 89
waterfronts, 29, 123, 133
waxworks, 89
wildlife attractions, 87
wildlife parks, 67
World Heritage Convention, 158, 160
World Heritage Sites, 158–62
World Tourism Organization, 51
Wrexham, 189

Wroclaw, 43; Cathedral of St. John the Baptist, 81; Ostrów Tumski, 58, 81

York, 15, 102

zoos, 87

# About the Author

**Dr. John Ebejer** is Senior Lecturer in the Institute for Tourism, Travel and Culture of the University of Malta. He has authored several journal articles and book chapters mostly on the subjects of tourism, historic areas and urban regeneration. He has also delivered lectures in universities in the UK, Finland, Austria and Italy. Before dedicating himself full-time to academia, he worked for many years as an architect, urban planner and tourism consultant. He was consultant to public sector agencies on public projects and on tourism product development. Ebejer has coordinated the implementation of numerous tourism-related projects. Earlier in his career, he was involved in urban planning initiatives, including the Tourism Master Plan for the Maltese Islands, various local plans and development briefs. Ebejer holds a first degree in Architecture from the University of Malta, a Masters degree in Urban Planning from the University of Sheffield and a PhD in Tourism from the University of Westminster.

www.ingramcontent.com/pod-product-compliance
Lightning Source LLC
Chambersburg PA
CBHW021847300426
44115CB00005B/45